AF574804

Mysteries of Nature

Great Mysteries

Aldus Books London

Mysteries of Nature

by Cathy Kilpatrick

Series Coordinator:	John Mason
Art Editor:	Grahame Dudley
Designer:	Adrian Williams
Editor:	Nina Shandloff
Research:	Marian Pullen, Frances Vargo
Series Consultant:	Beppie Harrison

SBN 490 004342

First published in the United Kingdom
in 1979 by Aldus Books Limited
17 Conway Street London W1P 6BS
Printed and bound in Hong Kong by
Lee Fung-Asco Printers Ltd.

Introduction

How has our earth evolved over the millions of years since the beginning of time? Has the globe always looked as it does now, or did land and sea areas shift at some point? Is the earth stable now? These and other questions concerning both the earth's mysterious past and its inhabitants are the subject of this book. What really happened to the dinosaurs, and how does Charles Darwin's famous theory of evolution relate to them? Will we ever know for certain what the world of the dinosaurs was like? Which living organisms so closely resemble their ancestors that they are called prehistoric survivors? Some animals have special talents—for navigation, camouflage, architecture, or even for social organization. How do they acquire these skills, and how do they know when to use them? And perhaps most importantly, where does man fit into this scheme of things: if he could predict earthquakes, for example, would he be able to prevent them? How was man himself evolved? Some modern animal species are the direct result of man's interference in nature by the domestication and breeding of animals to suit his own convenience. When did this process begin? Finally, what must man learn about the fragile networks which interconnect living things and their environment so that he does not alter irreversibly the balance of nature on this planet?

Contents

Chapter 1
The Restless Earth

Earthquakes and volcanoes have fascinated and terrified human beings since the dawn of time. Primitive man thought a monster lived inside each volcano who, when angered, would cause an eruption in order to punish his enemies. Not surprisingly, people living in vulnerable areas developed rituals of sacrifices and prayers in an attempt to keep the monster contented and quiescent. But ironically it is through studying earthquakes and volcanic eruptions that modern scientists have discovered most of what we now know about what lies beneath the earth's crust. Over 150,000 earthquakes of varying degree are now recorded each year, and it is the patterns of their vibrations which give a picture of the mysterious and so far inaccessible interior of our restless earth.

On May 8, 1902, the 28,000 inhabitants of Saint-Pierre, on the lovely West Indian island of Martinique in the warm Caribbean Sea, were going about their daily tasks. Colorful tropical flowers bloomed in the lush vegetation while birds sang sweetly. Suddenly the entire population of the town, except for two people, was dead. The earth, once so calm and beautiful, had become restless. Described as an "act of God," the horrible disaster was caused by the violent volcanic eruption of Mount Pelée, situated six miles from the town. The victims died suddenly, within moments. This was because when the volcano erupted it pushed forth a glowing cloud of hot gases and incandescent particles which did not contain enough oxygen to support either burning or breathing. This cloud traveled at a speed of about 490 feet per second, and the air within it reached a temperature of 1450°F. The effect was devastating—tropical trees were uprooted as if they were matchsticks, while buildings were demolished by pieces of rock hurled like jet-propelled missiles from the volcano's interior. After the cloud had passed, the oxygen concentration in the air returned to normal, and the city was soon aflame. Those who survived the scorching temperatures were soon burned to death in the flames.

One of the two survivors was a prisoner jailed in a poorly-ventilated dungeon, with only a tiny open grate in the door.

Opposite: the eruption of the Kilauea volcano on the island of Hawaii in November 1959. This pit, called Kilauea-Iki, is adjacent to the large crater or caldera at the summit of Kilauea. At one point during the active period Kilauea-Iki contained a lake of lava 400 feet deep. The village of Kapoho, nearly 30 miles away, was destroyed by lava flows that eventually extended the shore of the island by as much as half a mile into the sea.

Terra Firma?

Sufficient heat entered to burn him severely, although his clothing did not catch fire. Buried for four days, he was eventually rescued by people from a neighboring village who had come to investigate. He was apparently later pardoned for his crimes after having survived such a catastrophe, and spent the rest of his life touring with a circus as the "Prisoner of Saint-Pierre."

The second survivor was a shoemaker who lived in a part of town on the fringe of the glowing cloud. He rushed into his house with a few others when he felt "the terrible wind, trembling earth and saw a darkened sky," but he was the only person in the house to survive, and managed to run away on burned and bleeding legs before the wave of fires started.

The Saint-Pierre disaster is only one of the hundreds of violent eruptions that testify to the dramatic power of those natural enemies of man, the earth's underground cauldrons, earthquakes, and volcanoes. The phrase "steady as a rock" is often used to describe a person who is strong and firm, because we tend to characterize the earth beneath us as permanent and stable. However, the frequency of volcanic eruptions and earthquakes in many parts of the world shows that this is not true. The earth is a restless, ever-changing place, and although these changes are often quite small, they become quite noticeable over long stretches of time.

Below: the ruins of Saint-Pierre after the 1902 eruption of Mount Pelée, which looms up behind the town. The volcano had been emitting ash and poisonous gas for nearly two weeks, causing small animals like cats and dogs to suffocate and drop dead in the streets. On May 8 a glowing cloud overflowed from a notch in the crater wall and traveled down the river valley toward the town, preceded by a rapidly moving surge of hot gas and suspended ash. Wisps of smoke still rose from the smoldering debris of Saint-Pierre on May 14, a week after the eruption.

Left: a rock formation in the Navajo reservation in Monument Valley, Arizona. Flat-topped, steep-sided rock mesas, buttes, chimneys, and pinnacles are common features of the landscape, carved by running water from horizontal layered beds of red sandstone. Desert rains, usually torrential downpours lasting only an hour or two, cause flash floods and short-lived streams which carry rock fragments and have great erosive power. These rock formations are the remnants.

The rock of the earth's surface is slowly changing by means of disintegration and decomposition. Various external processes combine to produce an overall effect called weathering, or *erosion*. Rain, snow, frost, ice, the waters of seas and rivers, evaporation, and temperature all play their part in the weathering process. Frost, for example, can actually break up rocks into smaller fragments. This occurs when water fills the small cavities and cracks in a rock and then freezes at night. The volume of the water increases by about 10 percent, great pressure is exerted and the rock can shatter.

Weathering also results from the action of chemicals. Running water, for example, can contain carbon dioxide derived from the atmosphere. At such times it acts like diluted acid and can change calcareous or chalky rocks such as limestone by dissolving and carrying them away in the form of calcium carbonate. Even animals help to further the process of weathering. Worms eat soil, extract organic goodness, and then deposit the finely ground waste particles at the surface of the soil. Burrowing animals such as rabbits and moles also help to loosen the surface material. Long-term weathering and erosion have resulted in some amazing natural wonders all over the world.

Although much is known about the earth's crust, what lay beneath it has been a mystery for centuries, and it is mainly through studying earthquakes and volcanic eruptions that scientists have begun to pierce the mysteries of the earth's inner depths. Many people associate volcanoes with earthquakes. Indeed, they are very often found in the same place. For example, there are plenty of both in Japan, New Zealand, and Mexico. However, volcanoes and earthquakes are actually two quite different aspects of the same overall geological structure that applies to all of the earth's crust, whether it is covered by sea or land.

Below: the effects of water erosion in the Badlands of South Dakota near the Red Deer river. Here also the rushing of flash floods through normally arid gulches, ravines, and arroyos has contoured the earth into intricate patterns of hills and ridges.

The Glow of Vesuvius

Right: *The Destruction of Pompeii and Herculaneum* by the early-19th-century British landscape painter John Martin. Mount Vesuvius erupted on August 24, 79 A.D. when the pressure of steam and gases inside the volcano rose high enough to blast off the peak of the mountain. The Roman commercial center of Pompeii was overwhelmed with hot, fuming ash while the smaller, mainly residential city of Herculaneum was buried under volcanic mud. A violent local earthquake had occurred 16 years before, and tremors had continued for several years. This warning of impending volcanic activity was not recognized as such by the local population. The rim of a huge volcanic crater survives on the northern side of Vesuvius as Mount Somma, and on the southern side a new cone has been formed since the eruption.

Vulcan at Work

Right: *Vulcan Forging the Thunderbolts of Jupiter* by the 17th-century Flemish painter Peter Paul Rubens. This Roman god of fire and patron of metalworkers, who was frequently portrayed as a blacksmith, came to be associated specifically with the volcanoes of Italy and Sicily, and from his name we acquired the word "volcano."

Volcanoes and earthquakes have figured in many myths and legends regarding the world to be found under our feet. Primitive man simply thought volcanoes were alive when they became active and brought destruction to his land and death to his family. Not surprisingly, he thought a terrible monster lived inside, whose wrath could be kept under control through offerings of food and gifts, sacrifices of animals and sometimes humans, and by deep prayers to the god. Various races have interpreted the monster myth in different forms. Primitive Indian tribes visualized the monster as a gigantic mole or a ferocious bear, while the early inhabitants of Chile believed it to be a huge whale. The ancient Japanese thought a spider of tremendous size and savagery lived inside each volcano, while the Indonesians believed that Hontobogo, a snake which held up the world with its body, dwelled within. If he moved the earth shook, and volcanic fire shot out of the top of the volcano, which terrified even the local gods. But was it so stupid for these primitive people to think that volcanoes were alive? Not really, for volcanoes are quite similar to living things in certain ways. The summit regions of some volcanoes, Mount Etna in Sicily and the Kilauea crater on Mauna Loa in Hawaii, for example, do tilt strongly when they are about to erupt, and they also look as if they are breathing because they inflate before erupting and deflate afterward.

Greek philosophers, who had thoughts on many aspects of the physical and nonphysical world, also studied volcanoes and their activities. Plato believed a river of fire ran inside the earth from which air escaped and thereby caused earthquakes and volcanic eruptions. His "Pyriphlegethon"—in classical mythology one of the five rivers surrounding the underworld—can be compared with *magma*, the molten material inside the volcano, and the escaping air can be recognized as the gases which are contained within the magma.

Aristotle had an alternative theory that ocean waves, breaking with force against the shore, compressed air into caves which led down into the earth. If the air then came into contact with sulfur or bitumen it would set them alight, and the flames, smoke, lava fragments, and ashes would escape through the vents provided by volcanoes. For more than 2000 years this "pneumatic" theory was the one accepted by most scientists.

The Romans thought gods such as Vulcan (a blacksmith) lived inside volcanoes. They did not realize, however, that volcanoes "slept," and that supposedly harmless, quiet ones could suddenly become dangerous. The eruption of Mount Vesuvius, on the shores of the Bay of Naples, around noon on August 24, 79 A.D. took them unawares. The Roman commercial center of Pompeii was buried under the volcanic mud along with the smaller, mainly residential city of Herculaneum. The eruption was recorded in detail by the writer and statesman Pliny the Younger, and his detailed eyewitness account is confirmed by analysis of the rock products of the eruption. Some passages in Pliny's letters show remarkable similarities to modern eyewitness accounts of glowing clouds, such as the one that enveloped Saint-Pierre in 1902. Extensive excavation at Pompeii has greatly increased archaeologists' knowledge of ancient Roman life at the time of the eruption.

Above: the modern city of Pompeii. Excavation was begun in 1748 and has continued up to the present day.

Below: bodies of victims of the Vesuvius eruption, 19 centuries old. The ash that covered the town formed a mold which hardened before the bodies decayed. More permanent casts have been made from these molds by filling them with plaster, allowing it to set, and then carefully chipping away the outer ashen covering.

Journey to the Earth's Center

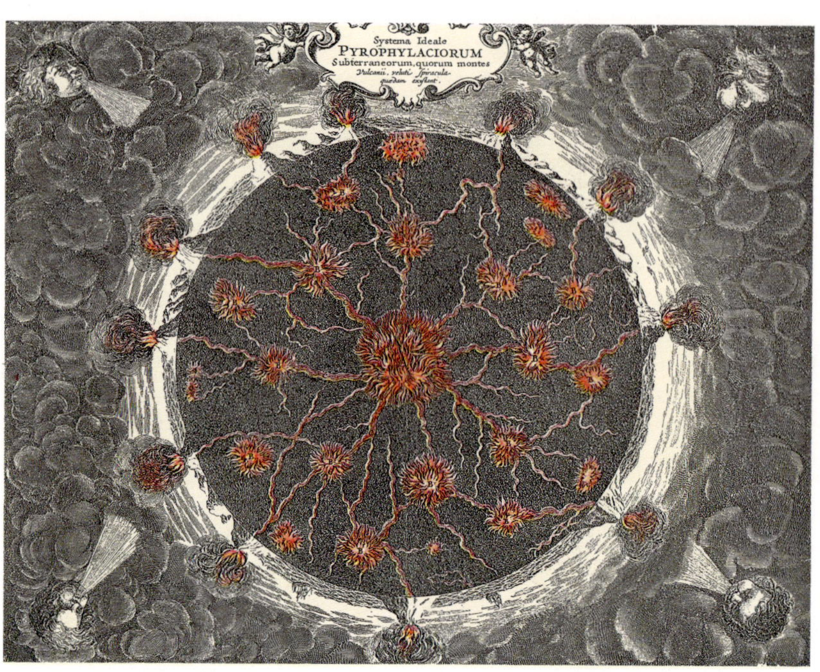

Right: Latin diagram of Aristotle's theory of the cause of volcanic eruptions. In his *Meteorics*, he suggested that both volcanoes and earthquakes were caused by winds trapped within the earth.

Through past centuries scientists and naturalists tried to discover the source of the heat which fed volcanoes. A theory of combustion such as Aristotle's was first widely accepted, but with the discovery of electricity in the 18th century it was thought that this might be the origin of volcanic energy. In the 19th century progress in the field of chemistry and increasing awareness of the heat-giving, or *exothermic*, reactions between water and metals like sodium, potassium, calcium, and iron led to the linking of volcanic origins with these types of reactions.

In this century volcanoes have been studied in great detail, and with the discovery of radioactivity several new hypotheses have been put forward. One theory suggested that rock inside the earth melted due to heat from the radioactivity of uranium, thorium, and potassium. Another said volcanoes were simply natural nuclear reactions, or atomic "bombs." In the 1960s an American chemist, Nobel laureate Harold C. Urey, suggested the startling theory that the planets had been formed by the clustering of meteorites. Although it attracted a great deal of support, this brilliant hypothesis was proved wrong when American astronauts brought back lumps of lunar rock as part of their moon exploration mission in 1969. Analysis of this lunar material showed that another theory, which claimed the earth had originated through heat, was correct. The lunar rocks were only 1 percent meteorite in origin, while the rest was made up of basalt magmas and igneous rock products, which occur in volcanic areas. On earth we can expect volcanoes to erupt and spurt forth magma until such time as the rocky crust of our planet has been sufficiently thickened to prevent any further escape of the magma within it. When this happens the volcanic process will come to an end.

Only in the last century or so has much knowledge concerning the interior of the earth been collected, primarily from the analysis and recording of volcanic eruptions and earthquake tremors. Even mining for minerals from "deep inside the earth," during which the earth's surface has been well explored, has resulted in penetration to the depth of only a few miles, compared to the 4000-mile radius of the earth. Describing the earth as an apple, the deepest drilling so far done by man below the surface of the earth would not approximate even the piercing of the skin of the apple! Therefore no direct observation has yet been possible of the inner earth, although many fascinating fictional descriptions have enriched stories and films such as the 19th-century French novelist Jules Verne's *A Journey to the Center of the Earth.*

Thus scientists, mainly in the fields of physics and mathematics, have used indirect methods involving precise measuring instruments to discover the physical properties of the earth's interior. In this way it was found that the area deep inside the earth has a much higher density than the earth's surface. But the study of earthquakes and the vibrations they generate, known as *seismology*, has provided the most significant evidence. Instruments which record the occurrence of earthquakes are called *seismoscopes* and are capable of detecting even the slightest tremblings of the earth, which are so minute that humans do not even feel them, and no damage results. Over 150,000 earthquakes are now

Above: an illustration from Jules Verne's story "A Hole Through the Earth," in which an attempt is made to dig a gigantic shaft many miles deep into the earth.

Left: scientists measuring the temperature of Sicily's Mount Etna at night. A rough guide to temperatures—which vary depending on how deep and how gaseous the lava is—near the surface is provided by the color of the molten lava. Records of Mount Etna's volcanic activity begin around 800 B.C. with a reference by Homer, and an eruption is described in the *Aeneid.*

Above: effect of the Los Angeles earthquake of February 1971.

Below (left to right): one recording station (a) receives surface waves first because it is within 60 miles of the focus or center of the earthquake. A more distant station (b) first receives the deeper waves refracted at the Moho (between yellow and red) because they make up for the longer route by their greater speed. (continued opposite)

recorded each year, and it is thought that if enough seismoscopes were installed over the entire surface of the earth the figure would reach well over one million.

When an earthquake occurs, vibrations are sent out as shock waves from the source region, or *focus*, up to the surface of the earth and also deep into the interior. This phenomenon can be compared to dropping a pebble into a still pool of water. The waves pass outward from the point of entry and also downward through the water to the bottom of the pool. If this pool were composed of three different liquids which had different densities,

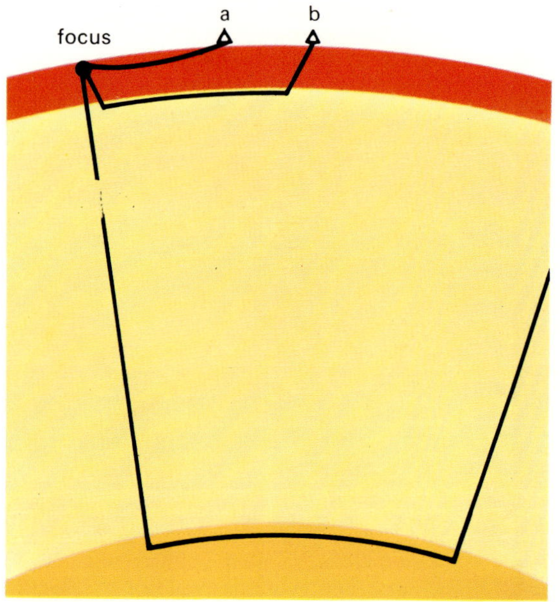

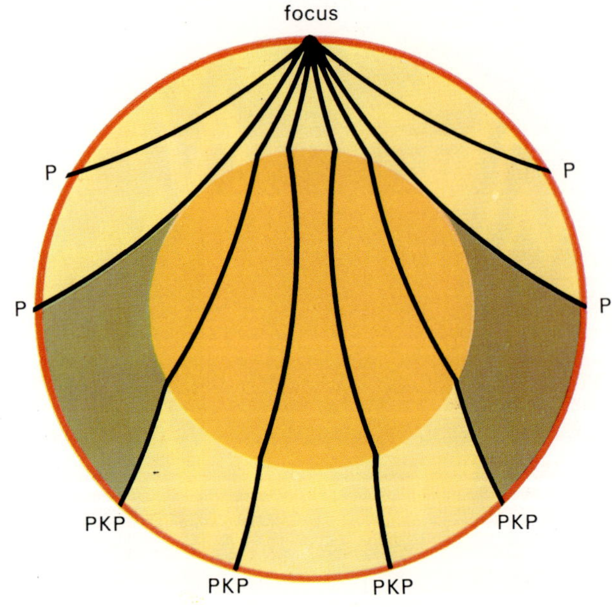

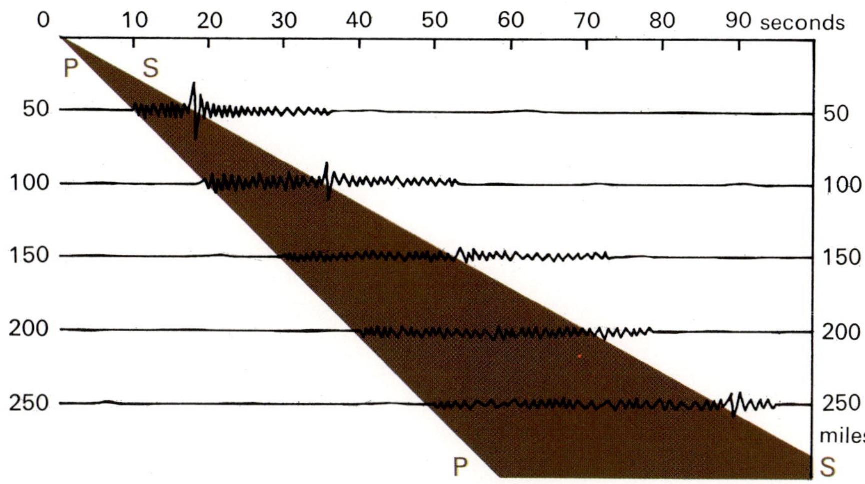

Earthquake Shock Waves

Left: diagram showing the approximate ratio between the distance and duration of travel for both P and S waves. P waves always travel faster than S waves. By distinguishing between P and S waves on a seismogram and measuring the interval between their times of arrival, the distance of the earthquake's focus from the recording station can be determined.

such as glycerine, water, and mercury instead of only water, then the course of the downward waves generated by the pebble would alter as they passed from one liquid to the next. This is what happens during an earthquake, since the earth is also composed of layers of different substances with different physical properties. Scientists, by studying the way in which earthquake waves pass through the earth and the time they take to do so, can deduce the presence of particular layers once they know certain information such as the compressibility or rigidity of the earth at various depths. This sort of study and analysis is important because an earthquake produces the only natural wave which passes through the earth in this way.

Scientists have found that earthquakes produce three main kinds of waves. The primary or compressional waves (P) are often termed the push waves since they make particles of rock move along their direction of travel, rather like a coiled spring when one end of it is touched. The waves, like the spring, are pushed back and forth and are similar to sound waves. These P waves move more quickly through the earth than the others, and so reach the recording station first. With secondary or shear

Below (continued from opposite page): On entering the core (orange), P waves are refracted downward towards the earth's center (waves that travel through the core are given the letter K). Because of this downward refraction, no PK waves can emerge from the core within the gray shadowed zone.

Some S waves are converted to P waves—and so gain speed—when they enter the core. Thus they are refracted upward (not downward) toward the mantle and eventually reach the surface within the shadowed zone. On leaving the core the SK waves may set up both P and S waves in the mantle and crust (SKP and SKS).

Many shock waves are reflected, like echoes, by the earth's surface and discontinuities. The PP wave has been reflected once from the surface. A P wave can also bounce off the core (PcP). The PKKP wave, having entered the core, has bounced off its boundary before reentering the mantle as a P wave elsewhere.

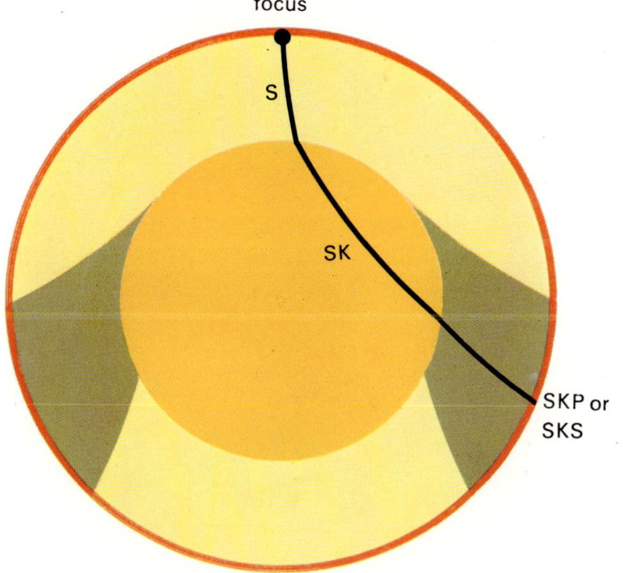

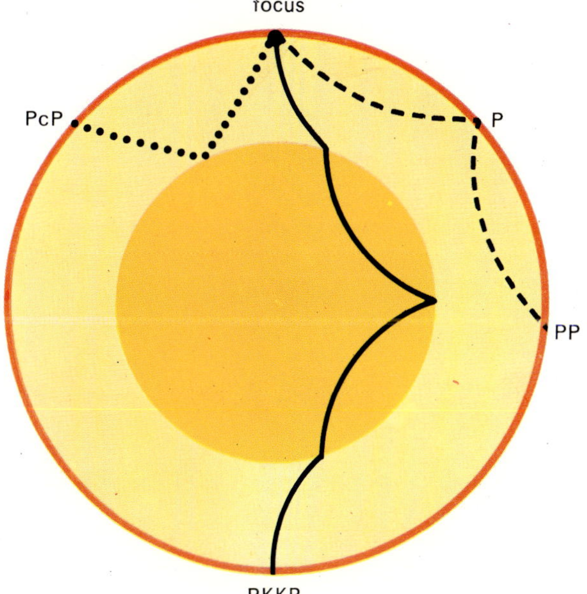

Drilling Through the Moho Layer

waves (S), particles transmitting the waves move back and forth at right angles to the direction in which the wave is traveling. They can be compared to the vibrations that are set up when a guitar string is plucked. In the earth P waves travel approximately 1.7 times as rapidly as S waves. The third type of earthquake waves are the surface waves, which travel only through the earth's crust. Their motion is like that of ocean waves, except that they are very low in height and quickly die out when they reach any depth below the ground surface.

Over years of observation seismologists have produced tables which show the relationship of distance to the time lapse between P and S wave arrival at a given point. For example, when an earthquake occurs 6000 miles away from a recording station, which is about the equivalent of one-quarter of the globe's circumference, the P waves take nearly 13 minutes to reach the seismograph while the S waves do not arrive until about 11 minutes later. The longitudinal surface waves are last to arrive.

If the earth were solid throughout, P and S waves would be expected to travel through the center of the earth in all possible directions. However, it has been found that there is a large area on the opposite side of the globe to the focus of an earthquake where S waves are not received, but P waves are. Why are S waves prevented from passing through the central region of the earth? Physicists found that S waves, like all transverse waves, cannot pass through liquids. This is indirect evidence for the theory that the earth's core is in the form of a dense "liquid."

Pioneer seismologists by 1914 had established several important features of the earth's interior. They had identified the three main components: the *crust*, the *mantle*, and the *core*. Earthquake waves were found to increase their speed suddenly at the boundary between the crust and the mantle, the middle layer. This boundary is popularly known as the Moho after the Yugoslavian seismologist A. Mohorovičić who discovered it by means of earthquake waves in 1909.

For many years thereafter it was impossible to verify the composition of the Moho or the mantle, as no samples were available from these depths. In the early 1960s the American government sponsored a deep drilling project aimed at reaching through the Moho and obtaining specimens of the mantle. Earthquake studies had revealed by this time that the Moho lay 20 to 40 miles beneath the heavy crustal rocks of continents, while in contrast, under the crust of the sea floor the Moho was found at

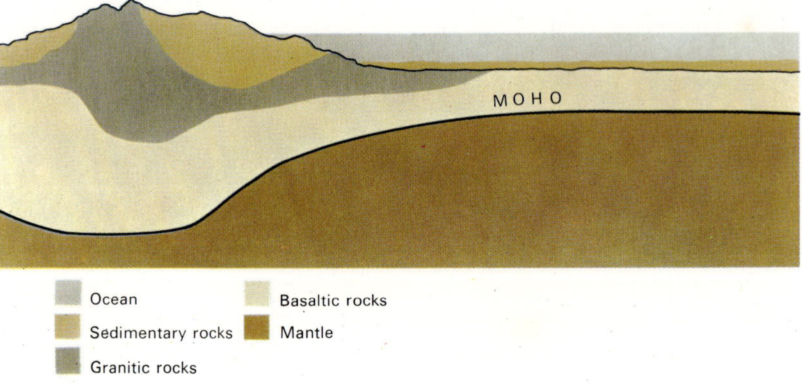

Right: the Moho, boundary between the earth's crust and mantle. The Moho may lie within three or four miles of the sediment-covered sea floor, but beneath the continents it is pressed downward to a depth of 20 to 35 miles by the weight of the crustal rocks.

much shallower depths, perhaps three or four miles down. The pilot drilling for the project was carried out in the eastern Pacific between the peninsula of Lower California and the island of Guadalupe in 1961. This went well and "cores" of the sediment covering the ocean floor were brought up. The main Upper Mantle Project, which inevitably was nicknamed Operation Mohole, involved drilling at a site 170 miles off Honolulu, Hawaii, in the northern Pacific. The problems of drilling through three miles of rock beneath five miles of water were very great. This was before the development of deep-sea oil rigs to meet the special requirements of the North Sea and other sites, and until this time man had seldom obtained samples from the ocean floor at deeper than 60 feet. Operation Mohole engineers lowered into the water a 3500-foot pipe tipped with a diamond-toothed, doughnut-shaped drill bit equipped to cut into the seabed. At various depths a 20-foot-long core barrel was dropped into the pipe to trap a cylindrical mass of material which was then retrieved using a cable and brought up to the drilling platform. A basalt core, for example, was brought up from about 600 feet beneath the sea floor. But Operation Mohole encountered many difficulties in regard to both drilling and politics. Although the initial budget was granted, in 1966 the project was abandoned when the United States Congress refused to sanction any further expenditure.

Until quite recently it was believed that the earth's core, with a diameter of 4300 miles, was a liquid mass of molten iron under immense pressure of about 24,500 tons per square inch, with an average density of 10.5. However, from the detailed study of

Above: the drill used by the Project Mohole drilling ship. Turned by a rotary table, it is lowered through a guide shoe in the ship's well. Sea water and drilling mud are pumped into the pipes to lubricate the drill bit, to which diamonds give a hard cutting edge. The drilled rock passes upward through the bit's center hole and will provide a continuous core from sea floor to mantle.

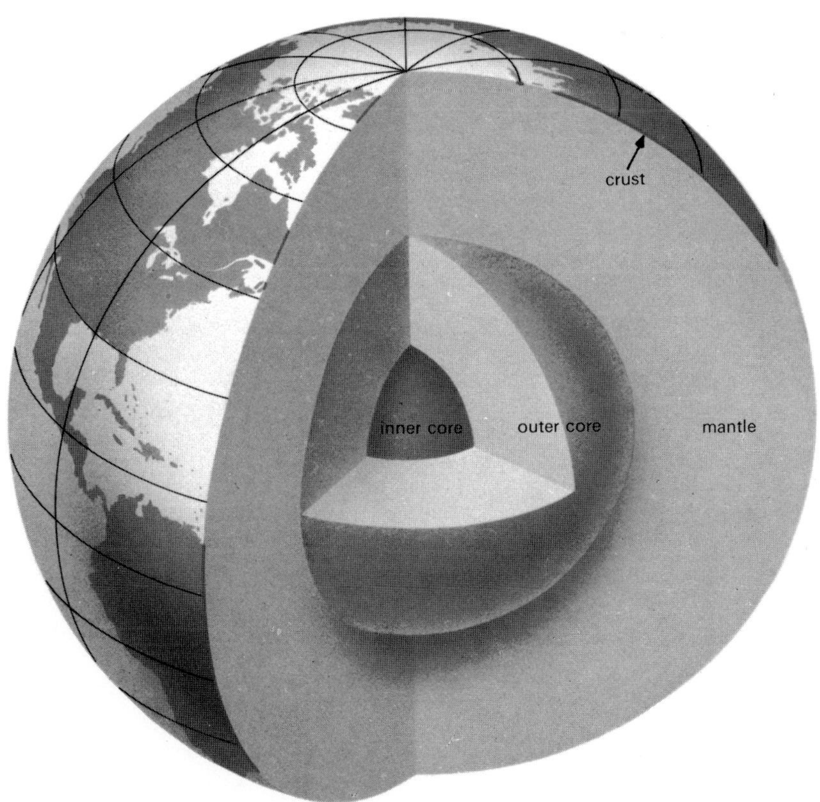

Left: cross-sectional drawing showing the structure of the earth. It consists of an inner core enclosed within three concentric layers —outer core, mantle, and crust. The earth has a volume of some 240,000 million cubic miles, and at its iron-rich center the pressure is about 20,000 tons per square inch. Only during the last couple of centuries has the developing science of geology—stimulated by industrial demands for fossil fuels, metals, and other minerals buried within the earth—begun to reveal the nature of the interior and of the immense forces that slowly but inexorably shape the crust.

What is the Richter Scale?

earthquakes in recent years, it is now believed that there is a central inner core, with a density of 16 to 17 and a diameter of about 1600 to 1700 miles. This solid mass is thought to be made up of an iron-nickel alloy similar to that found in meteorites.

Controversy still surrounds the composition of the outer core. It is thought to have a density too low for it to be a pure iron-nickel melt, so many scientists believe that something else must be mixed in with it. Certain geologists believe that substance to be silicon, while others have recently suggested that it is probably sulfur. If sulfur is present it is probably combined with potassium in various compounds, and as potassium is radioactive and an important heat-producing element, there are many fruitful lines of research into how it may relate to other elements within the earth.

In the continuing but not particularly successful effort to learn to predict earthquakes, thousands of seismoscope readings have been collected and interpreted. It has at least been possible to calculate the quantities of energy which are released as shock waves by earthquakes of various magnitudes. In 1935 a leading American geophysicist and seismologist, Charles F. Richter, introduced a scale of earthquake intensity which rates the quantity of energy released at the focus of an earthquake. The Richter scale consists of numbers ranging from 0 to 10 expressed as logarithms. The largest earthquakes observed between 1900 and 1950 registered 8.6 on the Richter scale, and released three million times as much energy as did the first atomic bomb. The earthquake in Chile in 1906 was one of these. The smallest detectable quakes register 0, equivalent to the shock of a 170-pound man jumping off a table. An earthquake is considered major if it registers 7 or above—and there are about 14 of these each year.

Below: seismograph belonging to the U.S. Geological Survey. Most seismographs work on the principle that an object will tend to remain at rest (inertia): a pendulum, hanging motionless from a piece of string held in the hand, will lag behind in its response to a horizontal movement of the hand. An earthquake wave moves the ground to which the machine (the "hand") is fixed, but the recording instrument itself (the "pendulum") is fitted with parts (the "string") that absorb most of the shock. Therefore it is the movement of the whole seismograph relative to the recorder that measures the intensity of the quake.

Below right: seismograph record from the Science Museum in London of the 1964 Alaskan earthquake. The needle almost broke in a frenzy of activity as it traced the tremors, the first of which were recorded at 4:45 A.M. on March 28—eight minutes after the disaster occurred in Alaska.

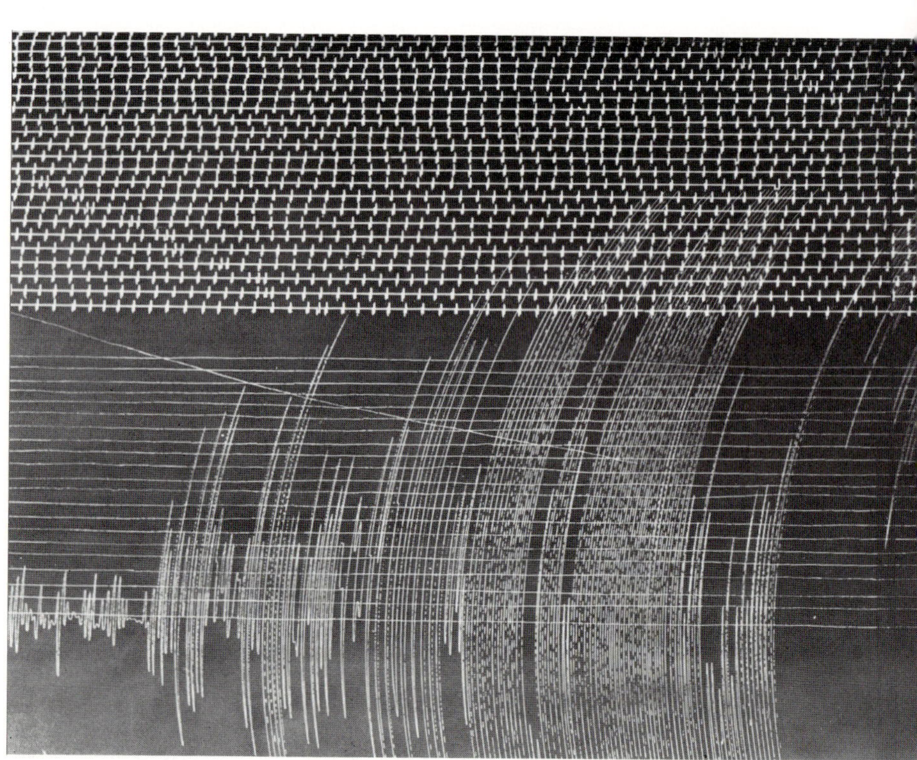

In 1956 Richter drew up a scale numbering from 1 to 12 called the modified Mercalli scale, in which intensity is described in terms of its effects on people at the earth's surface.

1 Not felt by people.
2 Felt by people at rest on upper floors of buildings.
3 Felt indoors: vibrations like a truck passing outside. Hanging objects swing. Duration of shock can be estimated.
4 Felt indoors: vibrations like a heavy object striking a wall. Stationary cars may rock; windows, doors, and dishes rattle.
5 Felt outdoors. Sleepers awakened. Liquids disturbed or spilled; wall-hung pictures move; doors swing.
6 Felt by all. People walk unsteadily. Plaster walls crack; furniture moved or overturned; windows, dishes, glassware broken. Trees and bushes shake or rustle.
7 Difficult to stand. Noticed by drivers of moving vehicles. Plaster, loose bricks, tiles fall; furniture and chimneys broken. Heavy bells chime. Waves on ponds.
8 Difficult to steer vehicles. Collapse of stucco and some masonry walls; fall, or twisting, of chimney stacks, towers, monuments. Frame houses move on foundations. Branches broken from trees. Cracks appear in wet ground.
9 General panic. Frame structures crack, underground pipes break, much masonry destroyed. Conspicuous cracks in ground; sand craters.
10 Most masonry and frame structures and foundations destroyed. Serious damage to dams and embankments; large landslides; sand and mud-flats shifted horizontally. Train rails bent slightly.
11 Train rails bent greatly; all underground pipelines out of service.
12 Damage nearly total. Large rock masses displaced; objects thrown into the air. Lines of sight and level distorted.

With the accumulation of all this knowledge, surely one of the most fascinating and useful aspects of a seismologist's work is to predict the occurrence of an earthquake in advance. The ability to announce that an earthquake or volcano of a certain magnitude will occur at a certain time and place would appease the fear and save the lives of many people as well as stimulating interest of scientists, philosophers, astrologers, and religious fanatics. However, although statistics reveal that some 14,000 lives are lost annually in these natural disasters, we are still a long way from accurate prediction, although there are, of course, ways to minimize the potential hazards.

Earthquakes cannot be predicted simply because we do not as yet fully understand how they are produced. The zones where earthquakes and volcanoes occur are well mapped, and it is known that the major earthquake belts are along island arcs and continental margins. But it is still a mystery why earthquakes and volcanic eruptions happen so abruptly and sporadically, so that prediction is almost impossible.

In recent years it has also become clear that man has inadvertently caused earthquakes through his intervention in certain geological processes. For example, underground nuclear

Above: structural damage caused by a recent earthquake at Gibelline, Sicily.

Above: chaos in a supermarket in Santa Rosa, California after a series of rolling earthquakes in 1969. Compared to the damage caused to buildings by the Sicilian earthquake (previous page), the effects of these tremors were minor.

test explosions in Nevada led to earth cavities being subsequently filled with water behind enormous dams, and fluids have also been injected into deep wells, all of which have caused increased earthquake activity in these areas. Thus a new line of research is suggested, that although we cannot predict earthquakes it may be possible to alter their characteristics in zones where major earthquakes occur.

It has been discovered that when man constructs and fills large dams, water accumulates behind them and earthquakes are frequently the result. After Lake Mead in Arizona was filled in behind the Hoover Dam in 1935, more than 600 local tremors occurred over the next 10 years. Seismologists now take great interest in dams and artificial lakes. Major earthquakes were also associated with the filling of a large artificial lake at Kremasti on the island of Rhodes in Greece, as well as the filling of the Kariba Dam in Rhodesia. On December 11, 1967 a disastrous earthquake hit Koyna Nagar in India, about 125 miles south of Bombay. Five years previously the huge Koyna Dam and a reservoir had been built, and up until this time there had been no earthquakes in the area. After 1962, however, a definite correlation was noticed between three factors—the height of the water in the

Right: Hoover Dam and Lake Mead, Arizona. The dam is in Black Canyon, on the Colorado river at the Arizona–Nevada border. Constructed between 1930 and 1936 as Boulder Dam, it was renamed in 1947 in honor of President Herbert Hoover. It is used for flood and silt control, power, irrigation, and domestic and industrial water supplies. It has a power capacity of 1,354,000 kilowatts.

Can People Cause Earthquakes?

reservoir (which varied with seasonal fluctuations in rainfall), the length of time the water was stored, and the seismic activity. Between August and December 1967 the water level was recorded at an all-time high, finally helping to produce the 1967 earthquake.

More evidence that man has been causing earthquakes comes from Denver, Colorado. Here a disposal well for waste fluids was drilled to a depth of more than 1000 feet into crystalline rock near Denver, for the use of the U.S. Army Corps of Engineers. Injection of fluid began on March 8, 1962, with the result that a striking relationship was observed between the volume of fluid injected and the number of earthquakes which occurred. The fluid injections were stopped at the beginning of 1966, but 1967 still showed a baffling increase in earthquake activity, which seemed to rule out simple correlation. However, scientists eventually formulated a theory to explain the continuing incidence of earthquakes. They suggested that injected fluids exert great pressure on faults in the rock beneath the earth, and this pressure causes earthquakes when the faults, no longer able to support the pressure, give way. When the injection ceases it seems that, although there is a rapid reduction in pressure at that point, the pressure continues to build up at points farther away. Large cracks at great distances from the fluid well are reactivated, although the smaller cracks near the well become inactive.

Left: the San Andreas Fault in California. Earthquakes are caused when rock strata fracture along a line of weakness, or fault, in response to the gradual build-up of pressure. The fault line is usually within the earth, but in some cases it is visible at the surface, as here. Over 180 million years the sides of this fault have moved horizontally more than 200 miles in relation to one another. The main factor has been the northward movement of the western side.

In some cases oil has been obtained by injecting water under great pressure into an oil field in order to force the oil to the surface. In the Rangley oil field in Colorado it has been found that earthquakes tend to occur where the fluid injection process generates the largest increase in fluid pressure.

Scientists are now studying the possibility of injecting fluids under pressure into the earth in areas of high earthquake activity in order to release smaller, less destructive earthquakes of lower magnitudes. Research toward earthquake control and eventual accurate prediction is continuing and seems well worth the effort if lives and cities might be saved. Within a comparatively short space of time, geologists are confident that they will be able to predict and lessen the effect of earthquakes and volcanoes.

AN NVS
IN PRINCIPIO CREAVIT DS CELV ET TERRAM MARE ET OMNIA QVAE IN EIS SVNT
ET VIDIT DS CVNCTA QVE FECERAT ET ERANT
DIXIT QVOQVE DS FIAT LVX ET FACTA E LVX
LVX
VBI DIVIDIT DS AQVAS AB AQVIS
VOLATILIA CELI
CVORAS SETIVDAS
IV DEI
IV DAS

Chapter 2
Origins of the Earth

Today it is more or less accepted that the earth is about 4600 million years old. But how it was formed is still just as much a subject of speculation as centuries ago when myths and legends were the only ways men could explain how their world began. In the 20th century, concepts of condensation and continental drift have largely replaced the seven-day creation as recounted in the Book of Gencsis, and the original single land mass christened Pangea may become the modern alternative to the Garden of Eden. Magnetism and convection currents are among the forces which geologists now believe have helped determine the geography of the earth, not only in the past but even now, as a continuing process beneath our feet.

God-fearing man has always been impressed by the account of the world's creation as given in the Book of Genesis in the Bible. Heaven and earth were reported to have been created within seven days: "In the beginning . . . the earth was without form and void; and darkness was upon the face of the deep. . . . And God said, 'Let there be light': and there was light . . . and God divided the light from darkness."

That was "Day One." On the second day God created a firmament, Heaven, to divide the waters below from those above it. On the third day the dry land appeared and vegetation clothed it. The fourth day brought the sun, the moon, and the stars, while marine animals and birds arrived the next day. The sixth day is significant both as the day land animals came into being and also because God made man in his own image as a separate act of creation. The seventh day was the day God rested.

It is perhaps noticeable that the inside of the earth, including ores and minerals such as coal or gold, is omitted in the seven-day saga, although gold and other semiprecious stones are noted in Genesis 2. Geological theories and interpretation for centuries conformed to the Bible's version of the creation of the dry land, also taking into account the description of the Flood which was

Opposite: medieval Spanish tapestry, in effect a calendar, showing the origins of the earth as described in the Bible. It shows the cycles of the year with God at the center, surrounded by representatives of the seven days of creation as described in the Book of Genesis. At the lower left the sun rides its chariot across the sky. Around the borders of the tapestry are depicted various months or signs of the zodiac, which are symbolized by the activities considered most characteristic of them.

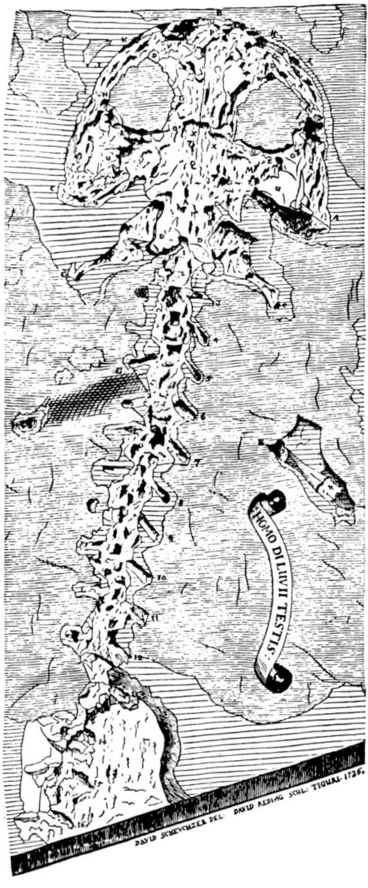

Above: fossil giant salamander found near Lake Constance in 1726 by a Swiss doctor, Johann Jakob Scheuchzer, who described it as "the bony skeleton of one of those infamous men whose sins brought upon the world the dire misfortune of the Deluge." Scheuchzer believed all fossils were the result of a catastrophic biblical flood and he decided that *Homo diluvii testis* had died out in 2306 B.C. precisely.

Below: visualization of the Hindu idea of the earth's position. The tortoise is the symbol of force and creative power. Here it rests on the great serpent, emblem of eternity. Seen gleaming at the summit of Mount Meru is a triangle (creation).

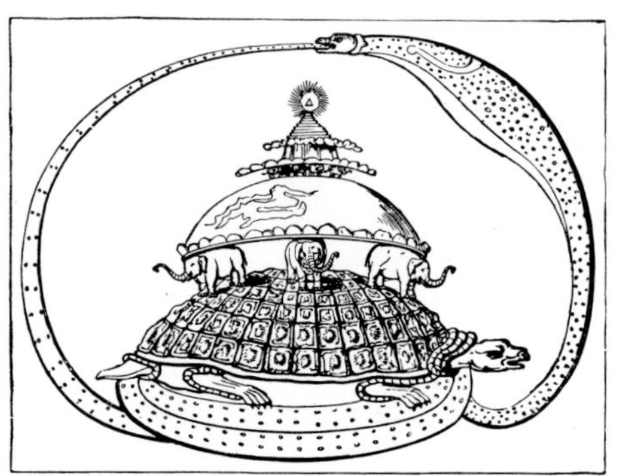

said to have destroyed most of mankind and almost all of the animals. The only survivors were Noah, his family, and the animals taken on board the Ark with them.

Many other religions offer different but just as fantastic stories of the earth's existence. The Egyptians believed their god of the air, Shu, parted the sky goddess Nut from the earth god Geb. Later, Greek mythology paints a similar picture: the sky is kept from falling to the earth because it is held on the shoulders of the giant Atlas. The Hindus have a very strange theory that the earth is supported on the back of four elephants, who in turn stand atop the shell of a tortoise that swims in a boundless sea. The theories all suggest that the earth is a flat plane in the center of the universe, and that the entire heavens revolve around it once a day.

Aristotle (384–322 B.C.), the Greek philosopher, attacked the "flat earth" theory, but it was not until explorers such as Christopher Columbus sailed "over the edge" of the so-called flat earth that the theory was proved wrong. Although there are many mysteries about our planet which have yet to be answered, we do now know a great deal about the earth.

Through centuries of studying fossils, rocks, and the earth's make-up, most scientists have come to agree that the earth is much older by millions of years than was previously thought. Today the age of the earth is more or less accepted as 4600 million years. At its "birth" the earth's surface ceased to be a molten, fiery chaos, and the continents which now cover our planet began to amass at a very slow rate. There was certainly no life on earth at this time, and it was not until millions upon millions of years later that conditions were right for the beginning of life. Life was probably triggered off about 3300 million years ago, but we have no fossil records from this period for proof as the life was only microscopic in size and had no hard parts to fossilize. In fact, until about 600 million years ago fossil remains did not exist. A vast period of some 2700 million years passed between the origin of life and the formation of the first fossil. Although we know that the earth is very, very old, no one yet knows exactly how and when the earth was created.

We now accept that the earth is part of a solar system which consists of nine planets and their moons, as well as assorted asteroids, comets, meteorites, and gases which revolve around a central star—our sun. The fact that all the planets circle the sun in the same direction, and that most also spin in the same direction and within the same plane, suggests that they were all formed at the same time and are of a similar age. Most scientists favor the concept of *condensation* to explain how the solar system was formed, and under this heading are three main theories. The solar system could have condensed from a vast revolving gas cloud which gradually cooled down and contracted as a result of its own gravity pull. The central mass remained hot enough for thermonuclear reactions to take place, thus forming the sun, a new star. The surrounding condensation formed the planets, which were only hot enough to keep their rocky interiors molten. A second, more remote possibility is that the planets and moons condensed from a streamer of gas and dust, torn off an already-existing sun by a passing star. A relatively new theory is that a

How was the Earth Formed?

disk of gas and dust, ejected from a huge, rapidly-rotating sun, condensed to form the planets. The larger, more remote planets such as Jupiter and Saturn were the only ones to retain large amounts of gas from the original condensation process. The inner planets such as Mars, the earth, Venus, and Mercury were stripped to their bare and rocky cores.

For centuries little thought was given to the question of why the continents and seas are shaped the way they are. It was generally accepted that they had always been there, in the same position, ever since the earth came into existence. Leonardo da Vinci (1452–1519) found fossil sea shells on Italian mountainsides and realized that the environment must have been different thousands of years before, as he wrote, "Above the plains of Italy where flocks of birds are flying today, fishes were once moving in large shoals." However, it was not until the late 18th and early 19th centuries that geologists began to make some sense out of the structure of the earth. Some scientists adopted the idea that some of the earth's hills were the remains of once-fiery volcanoes. There were, as usual, many doubters. Abraham Werner (1750–1817), a German geologist, stated that volcanic rocks had formed from the crystallization of sea water. In 1785 James Hutton, an Edinburgh doctor and geologist, made the astounding statement that the face of the earth had been shaped over long periods of time by processes of accumulation, heating, folding, and erosion. He also said these processes were still occurring. Most people, including scientists, did not accept his theory of *uniformitarianism*, and the same theory was advanced again years

Below: engraving from James Hutton's *Theory of the Earth*, published in 1785, in which he advanced his theory of uniformitarianism. It shows *unconformities* —horizontal rock strata resting on steeply inclined ones. According to Hutton such features were due not to sudden cataclysms but to very gradual folding movements allied to erosion.

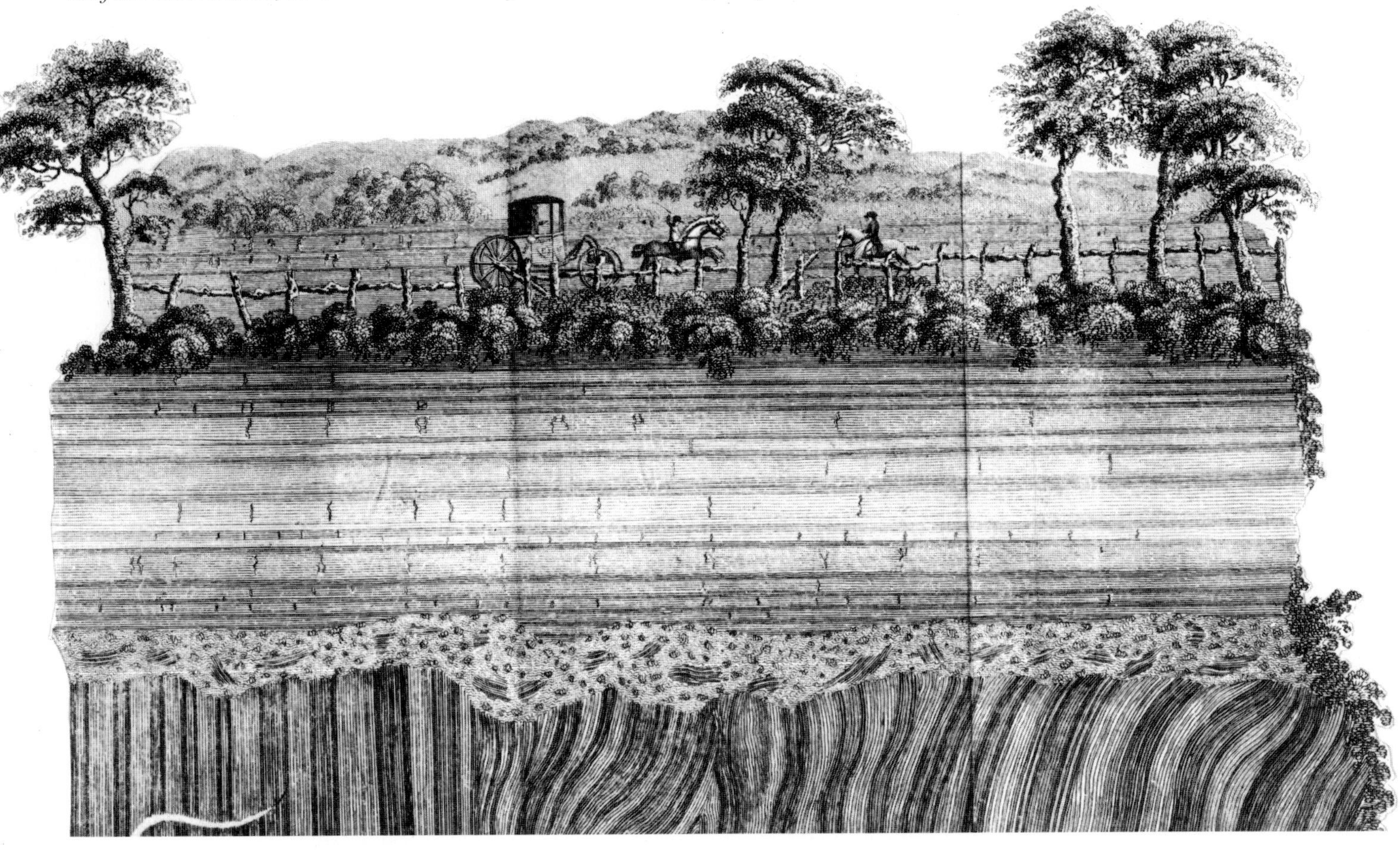

The Mystery of Pangea

later by another Scot, Sir Charles Lyell, who was born in 1797, the year Hutton died. Geologists had begun in the 19th century to collect vast numbers of fossils and rock samples from many parts of the world. Making use of this information, Lyell advanced the view that the earth's outer crust had been formed by the disintegrating action of erosion, followed by the resulting small particles being washed into the sea and then consolidated into new or secondary rock by the enormous pressure of water on the seabed. Convulsions beneath the crust caused the elevation of some areas and the depression of others, and those thrust above the surface were again acted upon by erosion.

But it was only in this century that the fantastic theory was suggested that the continents we live on are actually floating land masses, which have moved across the face of the earth during its long and restless history. Although Francis Bacon is sometimes quoted as having commented in 1620 on the similarity of shape of the coastlines on either side of the Atlantic Ocean, recent checks by N. A. Hupke of Princeton University in New Jersey turned up nothing in writing to suggest he had had these thoughts so far ahead of his time. It was a German theologian by the name of Theodor Lilienthal who cleverly interpreted the words from Genesis, Chapter 10, verse 25: ". . . the name of the one was Phaleg, because in his days the earth was divided," to mean that after the Flood the land had broken up. He also noted that some facing coasts, although separated by water, had coastlines which "matched" as though parts of a jigsaw puzzle—southwestern Africa and eastern South America, for example. By the close of the 19th century the geology of the southern continents had been quite well studied.

Below: Murchison River Gorge in Australia, showing its multicolored sandstone. Each bed was once a separate layer of sand spread by currents and deposited on the bottom of a body of water. Later the particles in the loose layers were cemented together by films of mineral matter from the water. Both the particles that settled from the water and the precipitated cement are sediments, which are the components of sedimentary rock. The color of sandstone, which ranges from grayish white through shades of yellow, brown, and red to very dark gray, are primarily due to the composition of the minerals in the cement.

In 1912 this amazing idea of moving continents, which is usually credited to a German, Alfred Wegener, was proposed. However, two Americans, Frank B. Taylor and Howard B. Baker, had already begun independently to outline their own ideas on *continental drift*. Taylor used the concept in 1908 to explain the origin of modern mountain systems, while Baker published a series of articles between 1911 and 1928 in which he suggested that the continents had been joined in such a way that modern mountain chains had formed continuous structures stretching from one continent to the next. Wegener, considered the pioneer of continental drift, was qualified as an astronomer and meteorologist, as well as being a competent polar explorer and balloonist. One of those increasingly rare scientists who delve into other branches of science in order to widen their own knowledge, Wegener studied geology and arrived at some revolutionary ideas, which he published in a slender book, *The Origin of Continents and Oceans*, in 1915. It was not until a third edition was translated and published in 1924 that his theory began to attract attention. Wegener suggested that all of the modern continents had once been joined together in a single land mass, which he christened Pangea (Greek for "all lands"), until after the Carboniferous period some 280 million years ago. It then began to break up and the pieces drifted slowly apart. Since these pieces would have shared the same characteristic flora and fauna, his theory explained some of the stranger facts of animal and plant distribution previously accounted for by the Austrian geologist Eduard Suess' hypothesis of land bridges in 1904. Suess had suggested that until the beginning of the Cenozoic era 70 million years ago, the world had consisted of only two land masses—a northern continent including part of Europe, northern Asia, Greenland, and North America, and a southern one which he called Gondwanaland which included Africa, South America, Australia, New Zealand, peninsular India, and Antarctica. Today's Mediterranean he considered a remnant of the extensive Tethys Sea that separated the two supercontinents.

If the four modern southern continents had been connected by land, then the distribution of flightless birds such as the rhea of South America, ostrich of Africa, and emu and cassowary of Australasia would be explained. However, Wegener believed that his own theory would better account for the distribution of such animals as side-neck turtles and marsupials, which are shared by Australia and South America. There was no geological evidence for the existence of Suess' land bridges, but linked with Wegener's idea of Pangea, they seemed a good way to explain anomalies observed in the animal and plant kingdoms. Wegener's theory contended that during the Cretaceous period around 100 million years ago, South America and Africa had effectively split apart, but that it was not until the present Quaternary period, which began 1 million years ago, that North America and Europe, or South America and Antarctica, finally separated. Australia had already severed its connection to Antarctica during the Eocene epoch, about 50 million years ago.

Vigorous counterarguments were advanced against Wegener, and geologists tended to look upon his ideas as those of a cranky meteorologist, unwilling to accept his deep understanding of

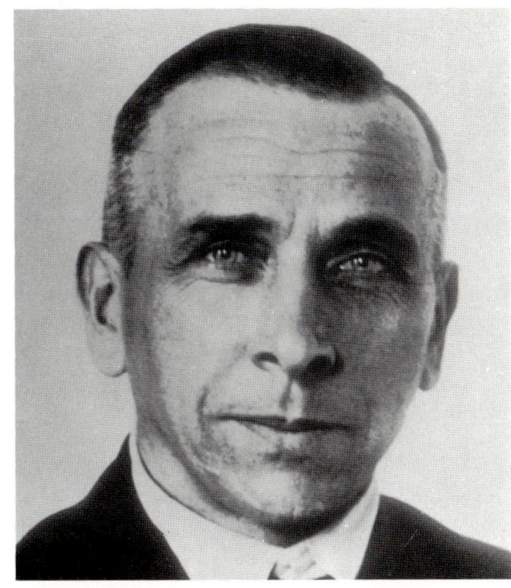

Above: Alfred Wegener (1880–1930), who first insisted that all the continents had once belonged to one gigantic land mass which he christened Pangea. Wegener was also a pioneer of meteorology.

Below: maps illustrating three stages in Wegener's reconstruction of the break-up of Gondwanaland, the original southern hemisphere supercontinent. They show Africa (1); India (2); Australia (3); Antarctica (4); and South America (5). The oceans are colored gray, and continental areas formerly submerged beneath shallow seas are marked in green.

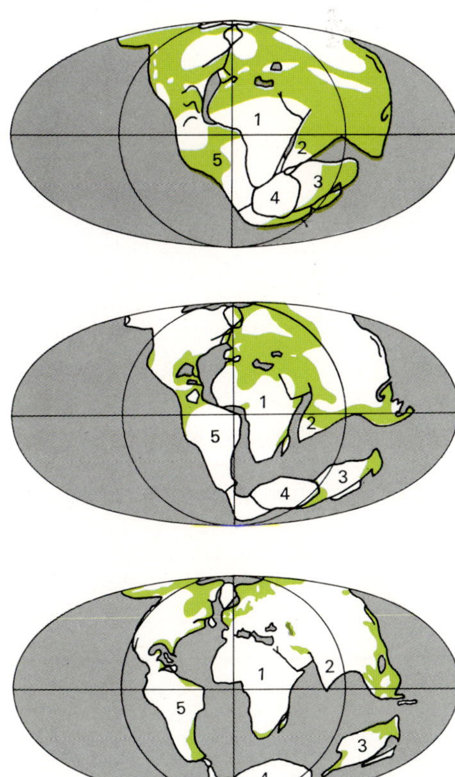

Geological Time Scale

Right: Precambrian fossil, one of many found in the Ediacara Hills in southern Australia since 1947. This is *Dickinsonia costata*, one of a variety of soft-bodied water creatures characterized as "jellyfish"; it has been preserved in sandstone. It may also be related to living worms. *Dickinsonia* ranged in length from a quarter of an inch up to 2 feet. Unfortunately no traces of eyes, legs or intestines have been preserved.

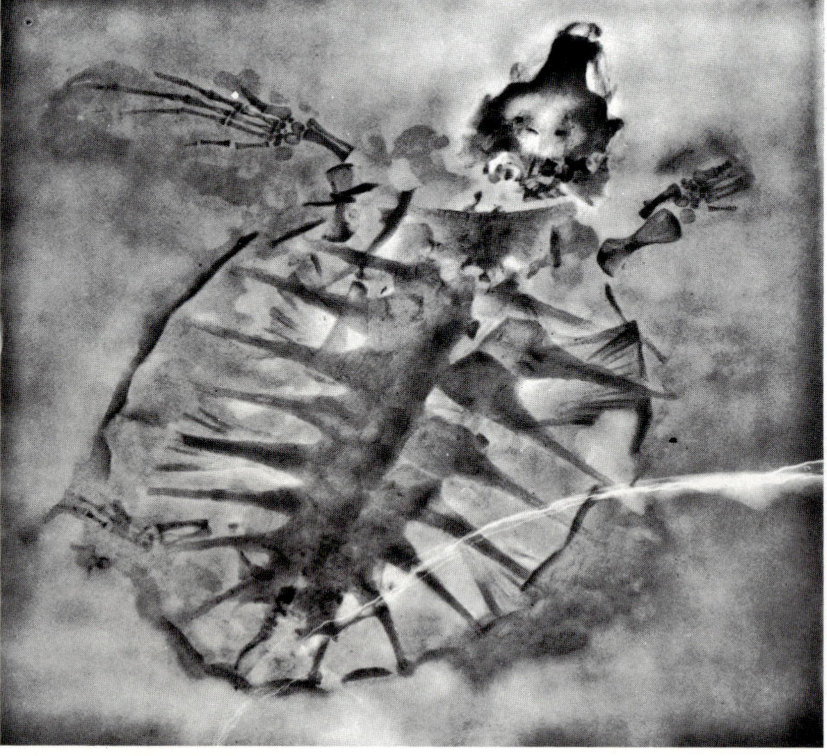

Above: X-ray photograph of *Glarichelys knorri*, a Mesozoic turtle. X-rays are used to discover whether a fossil is worth excavating and also provide an accurate guide to its placement within a bed of rock.

Right: table showing the geological eras, periods, and their respective time scales. The two fossils on this page are Precambrian—and therefore Cryptozoic—(top), and Mesozoic (bottom). The trilobite opposite is Cambrian and thus Paleozoic.

MAJOR GEOLOGICAL ERAS

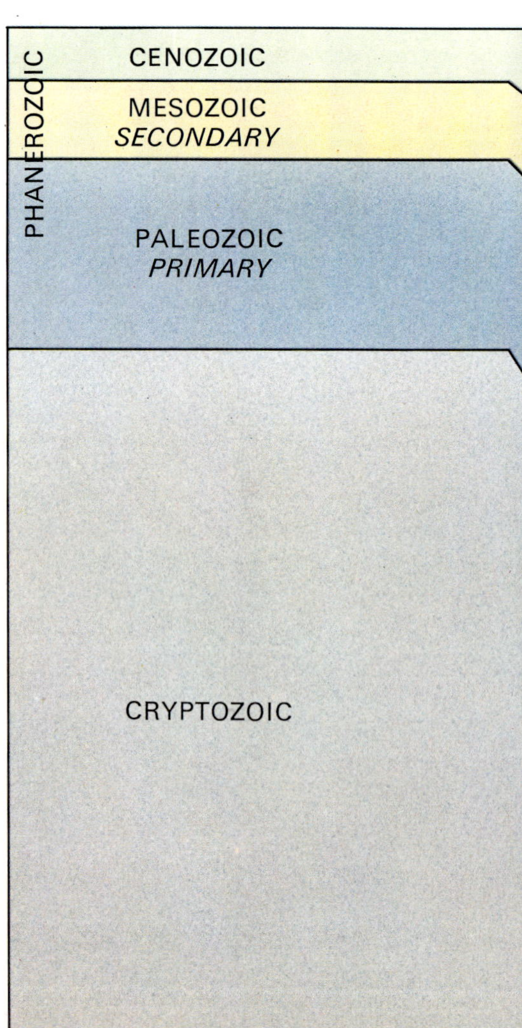

Left: a trilobite, a marine animal whose fossils first appear in Lower Cambrian rocks. Many species evolved during the Cambrian and Ordovician periods, and their fossils are useful for zoning and dating the rocks of these times. The last survivors became extinct in Permian times. Most trilobites ranged from nearly an inch to 4 inches long, but there were some giants of about 30 inches and many dwarfs only a tenth of an inch long. They had a hard outer shell, a number of jointed limbs, and a rigid plate for a head. Though primitive in some respects, the trilobites were highly evolved and very successful.

BEGINNINGS AND ENDINGS OF GEOLOGICAL PERIODS EXPRESSED IN MILLIONS OF YEARS

RELATIVE DURATIONS

	Period	Millions of years
QUATERNARY	HOLOCENE TO PLEISTOCENE	2-0
TERTIARY	PLIOCENE	7-2
	MIOCENE	26-7
	OLIGOCENE	38-26
	EOCENE	54-38
	PALEOCENE	65-54
	CRETACEOUS	136-65
	JURASSIC	190-136
	TRIASSIC	225-190
	PERMIAN	280-225
	PENNSYLVANIAN CARBONIFEROUS	310-280
	MISSISSIPPIAN	345-310
	DEVONIAN	395-345
	SILURIAN	435-395
	ORDOVICIAN	500-435
	CAMBRIAN	570-500

The oldest rocks so far dated are about 3300 million years old, but the Precambrian extends back to the Earth's formation, some 4500 million years ago. Despite attempts at world-wide correlation, Precambrian subdivisions still vary from one locality to another.

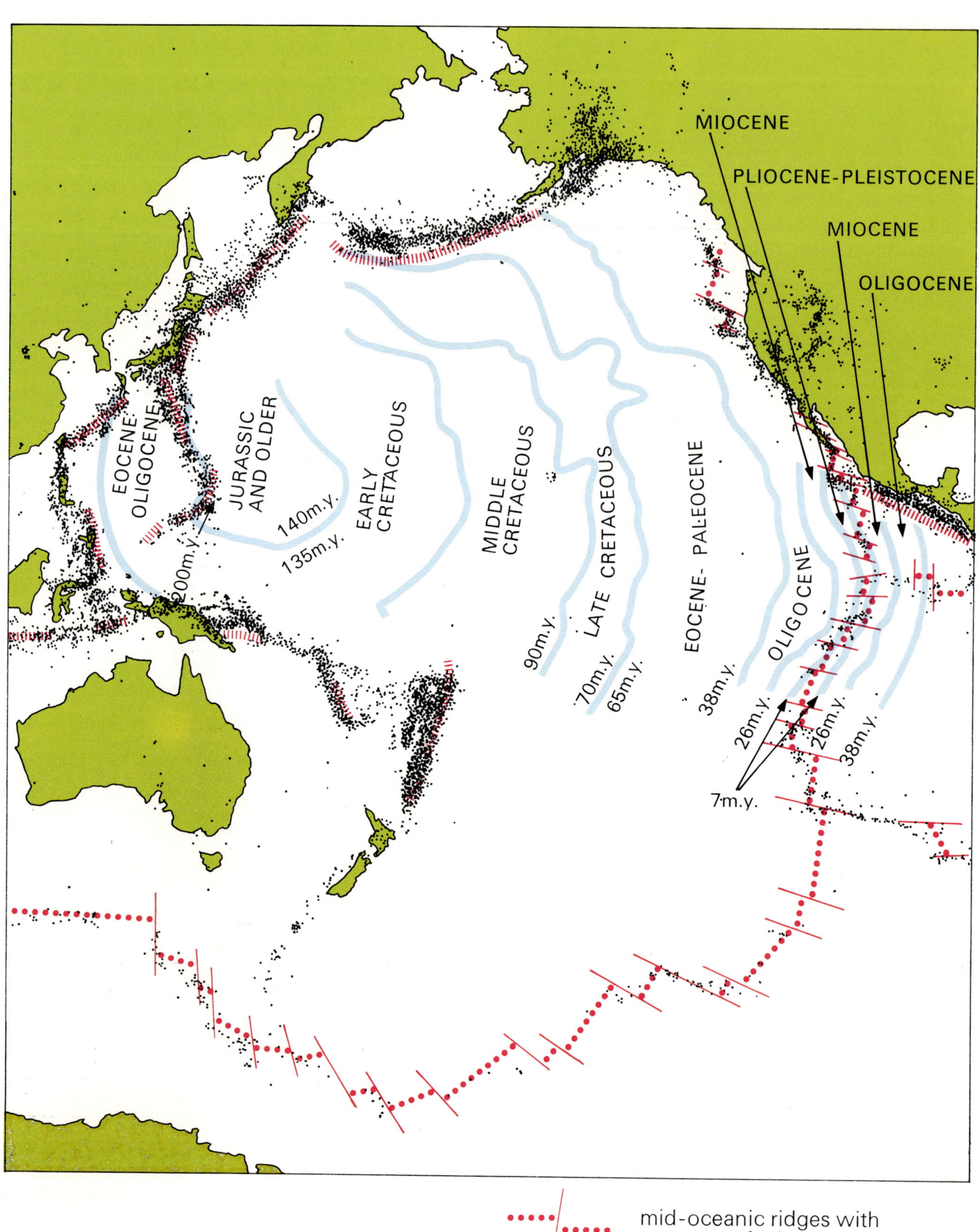

MIOCENE
PLIOCENE-PLEISTOCENE
MIOCENE
OLIGOCENE
EOCENE-OLIGOCENE
JURASSIC AND OLDER
EARLY CRETACEOUS
MIDDLE CRETACEOUS
LATE CRETACEOUS
EOCENE- PALEOCENE
OLIGOCENE
200m.y.
140m.y.
135m.y.
90m.y.
70m.y.
65m.y.
38m.y.
26m.y.
26m.y.
38m.y.
7m.y.
mid-oceanic ridges with transverse fault zones
active earthquake and volcanic zones
deep ocean trenches

geology. He died tragically on an expedition to Greenland in 1930, his theories already dismissed by the scientific community. Yet many, although not all, of his ideas have been proven correct in the last 10 to 20 years. He saw the break-up of the continents in much the same way that modern scientists interpret the earth's past, merely getting some of his dates wrong and overestimating the movement—his theory had the Atlantic Ocean widening some 10 times faster than is thought today.

But it might well be asked why, until the 1950s, few scientists bothered to consider Wegener's ideas at all. University students were taught for 50 years that Suess' land bridges explained the distribution of animals and plants through geological time, and Wegener's theory, if it were mentioned at all, was described as just a crazy hypothesis. It was only with World War II, when inventions were produced at an increased rate for war vessels to study the ocean floor and what lay under the sea, that many facts came to light.

During World War II special instruments were designed to search for underwater objects, from mines to submarines. However, this led to the use of these new instruments for recording and mapping the ocean bed. One major discovery was that the ocean floor and the crust beneath it are much younger geologically speaking, being in general less than 200 million years old, than the earth's crust which forms land continents. The underwater world is just as irregular in surface as the continental land masses, although under normal conditions the human eye is unable to see it. The deepest parts, such as the Mariana Trench in the North Pacific, which is over 36,000 feet deep, contrast with the higher parts such as sea mountains, ocean ridges, and finally volcanic mountains, which emerge above the surface of the water as volcanic islands such as Iceland or the Tristan da Cunha group.

There is now a great deal of evidence to back up the theory that the continents have drifted apart, and few skeptics remain. Although the coastlines of the two sides of the Atlantic Ocean do look similar and visually could be imagined to lock roughly together, it has since been realized that these are not the continental margins. The demarcations actually lie somewhere beneath the waves on the continental slopes, where the seabed inclines very steeply down to the deep ocean floor. Drawings of the contours of eastern South America and the West African continental edge were made in 1958 at a submarine depth of 1000 fathoms, and the similarities were very strong. In 1963 computers took over the work, although scientists had already worked out a theorem for fitting the continents together. It was found that the 500-fathom contour drawing made for a slightly better fit than the 1000-fathom contour. The excellent match between the submarine contours was very strong evidence that South America and Africa were once joined together. This idea was researched further by a South African geologist, Alexander du Toit, who looked in great detail at the soil structures of the three main areas overlapping on the map, including the Niger River delta and northeast Brazil. He suggested that the two continents had been joined together about 500 million years ago, so it seems the two continents must have drifted apart at some point between 500 and 50 million years ago.

The Evidence for Continental Drift

Opposite: the pattern and rate of sea-floor spreading in the Pacific Ocean as influenced by the movement of major continental plates. Mid-oceanic ridges (dotted lines) are offset by transverse fault zones (thin diagonal lines). The edges of continental masses are often rimmed by deep oceanic trenches (hatch marks). Active earthquake and volcanic zones (dots) usually fall along mid-ocean ridges or along deep trenches.

Below: deep core earth sample. Taking samples of the earth's crust by drilling enables modern geologists to make more accurate conclusions about its age and composition. This has made possible, for instance, the dating of the Pacific Ocean floor as shown in the diagram opposite.

Further evidence came from examining the rocks and fossils of the separate regions which would once have been connected. Although the evidence of matching shape and geological similarities could be merely coincidental, other facts were discovered which strengthened the case. "Wrong-latitude" evidence was compiled from the fact that various rocks and fossils, thought to require special climatic conditions for their development, were found in areas which seemed to have the wrong conditions. For example, coal is found quite near the South Pole, and yet to form in ages past it would have required a warm, subtropical climate. At the other extreme, the signs of huge continental ice sheets are evident in India. Did this not mean that land had moved to take up different positions during the course of time?

A major breakthrough for the theory of continental drift has come from a new branch of geophysics, called *paleomagnetism*. This is the study of the permanent magnetic lines of force shown in ancient rocks. We all know that a compass needle does not actually point to the North Pole but to the magnetic pole, as the entire earth is a huge magnet. The discovery of the earth's mag-

Below: map showing the distribution of certain fossil types common to the continents of South America, Africa, Antarctica, and Asia. *Lystrosaurus*, an extinct, sheep-sized reptile and so-called "index fossil," was among the first vertebrate land animals ever found in Antarctica in 1969. It was identical to those known to have lived in Africa, India, and China 180 to 225 million years ago; according to American paleontologist Edwin H. Colbert, this "indicates that Antarctica and southern Africa were joined along a broad front . . . making of them essentially a single land."

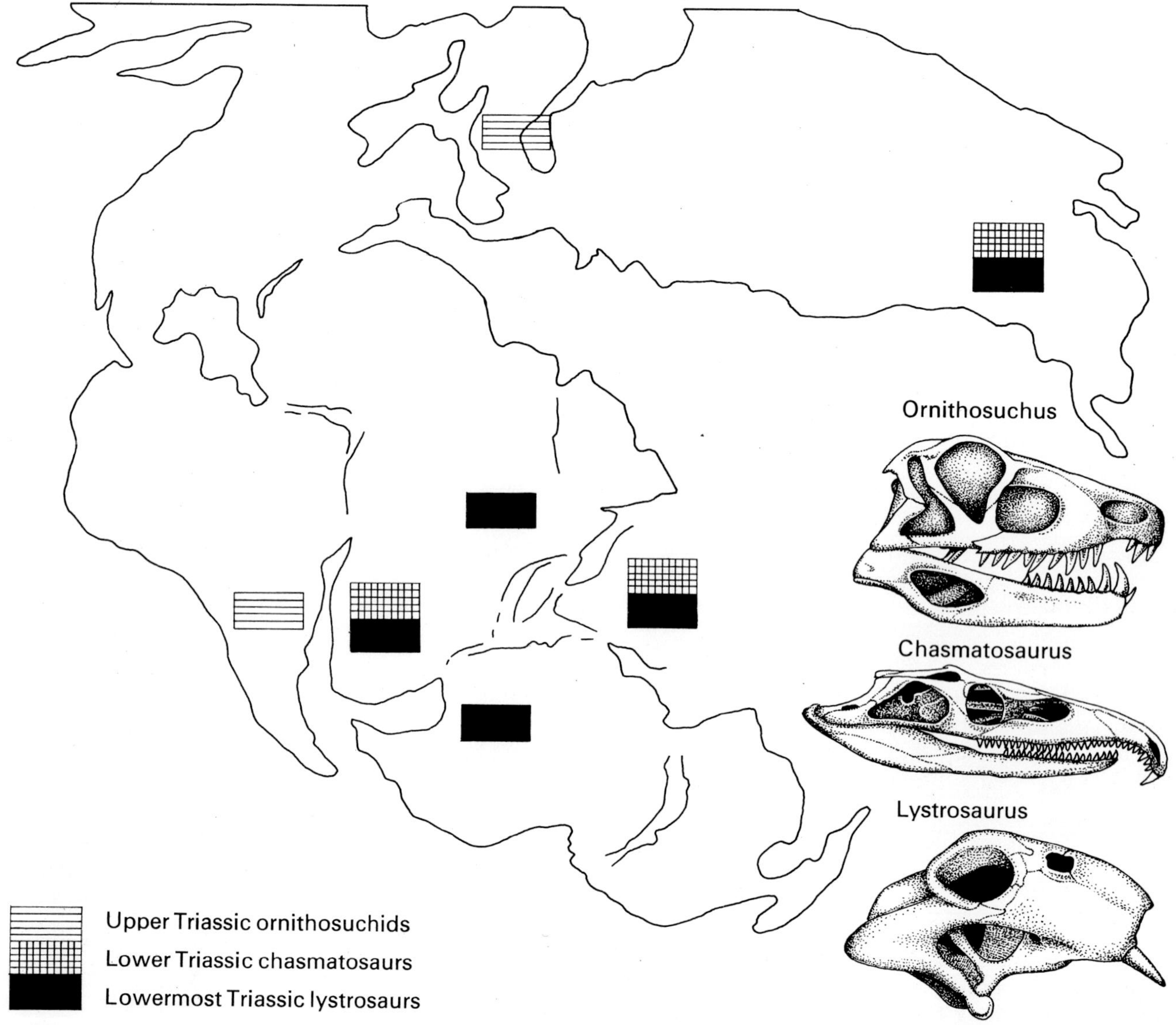

The Magnetism of the Earth

Left: formation in 1963 of the island of Surtsey, 11 miles southwest of Iceland, as the result of undersea volcanic activity along the North Atlantic ridge between two continental plates. Here huge columns of vapor rise from the sea at the point where it reaches the southwest side of Surtsey.

netism is credited to Sir William Gilbert, who was physician to Queen Elizabeth I of England. As long ago as 1600 he published a treatise called *De Magnete*, and in it he described how he used a tiny pivoted compass needle the size of a grain of barley to map the external magnetic field of the earth.

Simply explained, the earth's magnetic field is like a bar magnet placed at the center of the earth. The earth's magnetic axis lies at an angle of about 20 degrees from the center of the geographic axis. Even today scientists are not sure how these magnetic forces are produced or why the earth has a magnetic field. They do accept, however, that the answer is not the simple one that the center of the earth is made up mainly of iron. The terrific heat of the core would destroy the magnetism. A modern view is that the earth's magnetism can be explained by the *dynamo* theory. This suggests that the earth behaves rather like a dynamo in a power plant, in that the slow turning of the iron core sets up a magnetic field, thereby generating great electrical currents in the liquid mantle layer below the crust.

Ancient rocks show magnetic properties because when the hot lava or liquid magma which formed them cools, the various iron-bearing minerals in the rock became cool enough to retain their innate magnetic properties. When the rock hardens, the tiny magnetic particles are "frozen," immobilized in positions in line with the direction of the magnetic poles. As scientists extracted "fossil" magnetized rocks from the sea floor, they made strange and mystifying observations. Samples taken from mile upon mile out from both sides of the central sea ridges formed a pattern showing that the ancient rocks' magnetism alternated from north to south. It is now believed, partially based on this evidence, that over periods of thousands and even millions of years the north and south magnetic poles have wandered over a large area, probably at times in completely opposite directions. An overnight magnetic reversal would make a compass needle rotate 180 degrees so that the tip pointed in the opposite direction. However, scientists think the magnetic poles moved very slowly.

Defining Plate Tectonics

Another amazing feature which revealed itself from the study of ancient rocks was that the magnetic profiles across the width of the ocean on both sides of the ridge were very similar. If one is superimposed on the other they appear to be mirror images of each other. How did this come about? One explanation is that new ocean floor material is created by being pushed out from the middle of the oceanic ridges to each side. Using available data, scientists were able to suggest that in the Atlantic the sea floor has spread out from the central ridge at a rate of approximately three-quarters of an inch a year on each side. This means that the floor of the Atlantic is growing at a rate of about 1.5 inches a year. The spreading of the ocean floor seems to support the theory of continental drift, since it could have provided the force which caused the various continents to split off from each other in the first place.

There was more to follow in the mid-1960s when the term *plate*

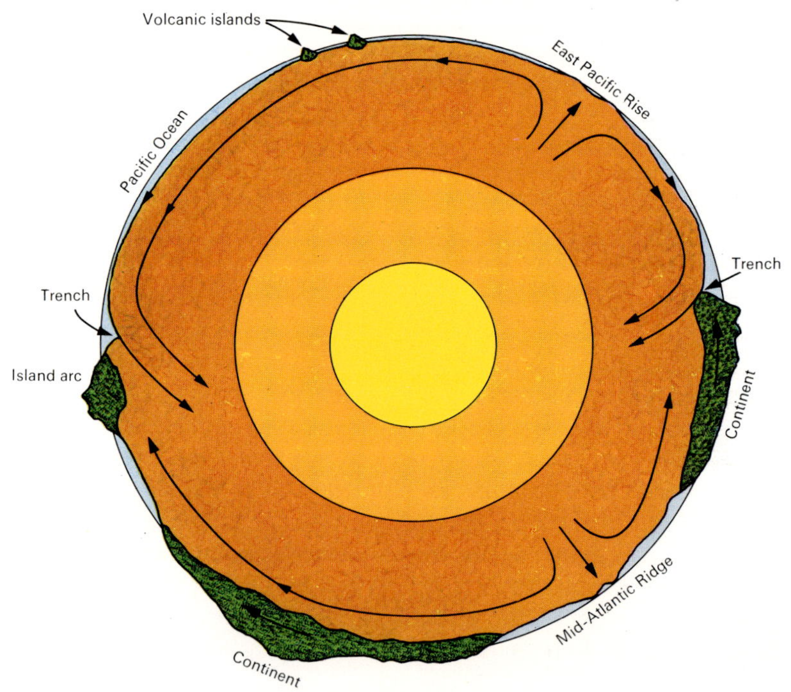

Right: schematic cross section of the earth based on the hypothesis of a spreading sea floor. The mid-oceanic ridges are outlets for the substances welling up from the mantle. It is precisely there that new suboceanic crust is born. This new ocean floor spreads out on both sides of the oceanic ridges and descends again into the mantle at the oceanic trenches. The speed of this "conveyor belt" process is estimated at over an inch per year. The ocean itself is several thousand million years old, but its floor has been continually renewing itself.

Below: the Canadian J. Tuzo Wilson, the first to use the term "plate" in 1965. He suggested that, given that the Atlantic Ocean was a gigantic rift and that centers of volcanic activity were localized at or near the center of the rift, the ages of the islands scattered throughout the Atlantic, all volcanic in origin, should increase the further they had migrated from the Mid-Atlantic Ridge.

tectonics (plate movements), describing a comprehensive system of theories, crept into the discussion of continental drift. In fact, today the new term is used in preference to continental drift. J. Tuzo Wilson of Toronto University was the first to use the term "plates" in a paper published in 1965. He considered those areas where the movements of the earth's crust were concentrated, and argued that they were connected to each other through a continuous network of moving belts which divided the surface of the earth into several large, rigid plates. A revolution in geological thinking was underway.

Two young researchers working independently in 1967–8, Dan McKenzie at Cambridge University in England and James Morgan at Princeton, discovered new formulae to fit the theory of plate tectonics. They used a combination of geometry and geophysics to show that the outer part of the earth's surface is divided into six major and some 15 or so minor plates. Four of

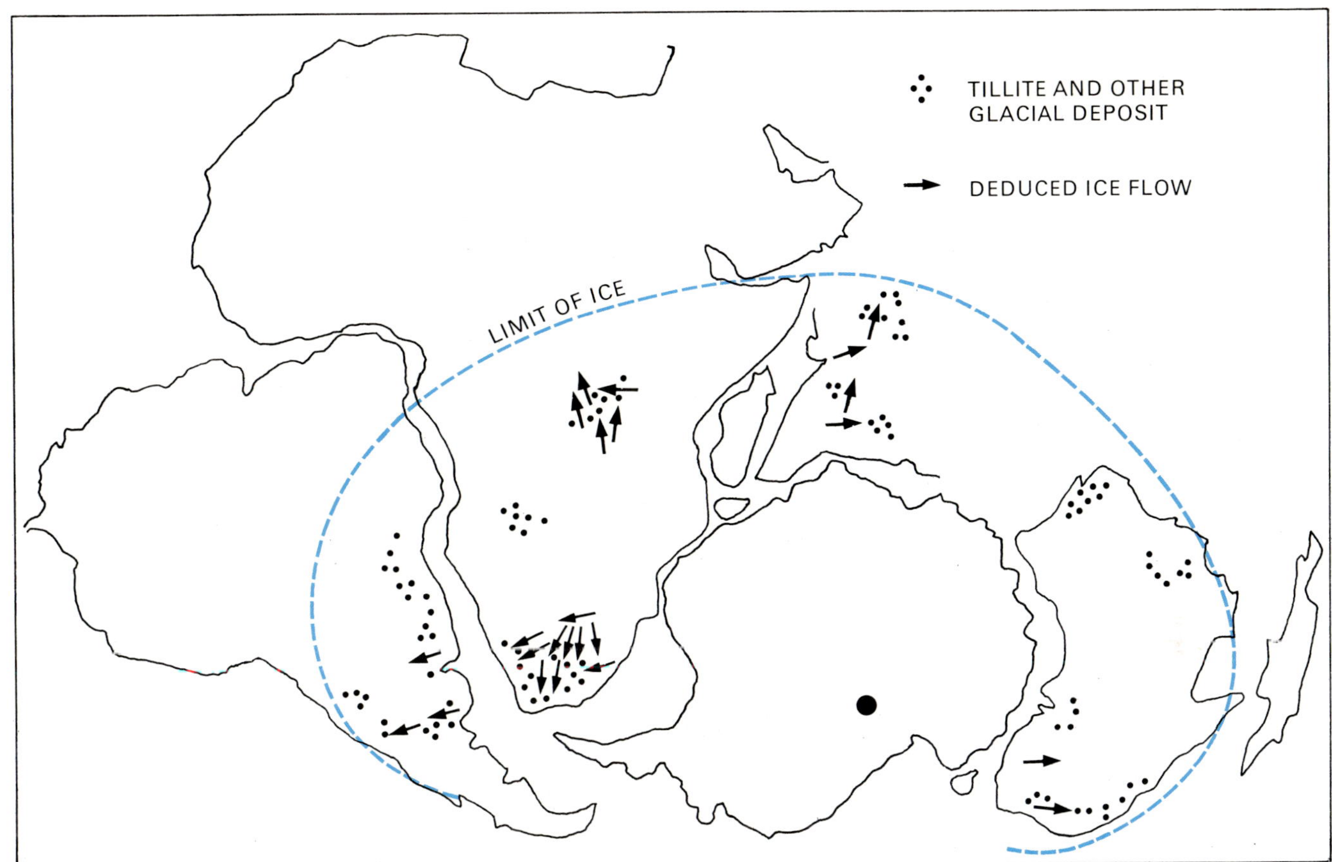

Above: diagram illustrating the direction of ice flow and the extent of glaciation over the supposed area of Gondwanaland in the southern hemisphere. The record of ancient refrigerations or ice ages, which have occurred at intervals during the earth's long history, consists of masses of tillite (consolidated till, a mixture of clay, sand, gravel, and boulders) in the midst of sedimentary or metamorphic rocks. What caused the glacial climates, however, is still a mystery.

the major plates carry continental land masses as well as ocean: the Africa plate (the continent of Africa), the Indian plate (India and Australia), the Eurasian plate (Europe and the bulk of Asia), and the American plate (North and South America and Greenland). The Antarctic is carried on the fifth plate, and largest of all is the Pacific plate, which has no continent floating on it. These plates move as rigid bodies, rotating about the globe in accordance with strict geometrical laws. The majority of geologists have finally been converted to the concept of plate tectonics in the last decade or so, but in the 1950s the "drifters" found it very hard going.

Further investigation has resulted in the discovery that the rigid plates can be deformed, but only where their motion in relation to each other causes two to collide or scrape. If two oceanic plates met at a deep sea trench, they would be destroyed by plunging down into the liquid mantle layer, where deep down in the hot interior they would melt and mix with the mantle from which they were originally formed. The destruction is compensated for by the creation of new plate material along the oceanic ridges of the globe. A simple comparison is that of a conveyer belt used to produce a continuous flow of objects. The ocean floor is produced at the oceanic ridges and then slowly pushed outward, receiving as it travels a thin layer or veneer of sediment from the water above. When occasionally part of the crust is forced down into the mantle it may be destroyed, but part of it will be "reborn," geologically speaking, as new continental crust by being shot out between plates as silica-rich lava from volcanoes in mountain ranges or island volcano arcs.

Churning Convection Currents

The massive floating continental plates are like gigantic floating rafts (Wegener compared them to icebergs), unable to sink because they are lighter than the denser rock of the earth's interior. Geologists have decided that when collisions take place the plates arch upward, breaking the granite crust into mountain ranges. An example is the formation of the great Himalayas between India and Tibet in Asia. There is still some doubt about the movement of the Indian subcontinent, but it appears that in the present Cenozoic era, which spans the last 70 million years, the Indian continental crust rammed into the thick sediment of the Tethys Sea and the land mass of Asia on the Eurasian plate to the north. The resulting crumpling and thrusting of the crust formed the Himalayas. Recently American geologist Bruce Heezen found evidence that the Maldive Ridge in the Indian Ocean was once a part of the Indian subcontinent, left behind as India traveled north in the early Eocene epoch some 50 million years ago. The mysterious East Indian Range, also called Ninety-East Ridge, in the northeastern Indian Ocean, could possibly be a 2500-mile skid mark left by the eastern edge of the continental plate. Some geologists believe that the Appalachians, a very ancient mountain range in North America, were formed from a continental collision at the close of the Paleozoic era around 250 million years ago. The Ural Mountains, which run north to south across the vast Eurasian continental plate, are thought to be the result of Asia and Europe colliding in the formation of the single continent of Pangea.

The theory of plate tectonics has undergone numerous tests to show whether it can explain various phenomena. One study examined the behavior of *faults*, which are the fractures at the surface of the earth, along which movement between adjacent land blocks occurs. It is known that when a fault is temporarily blocked, local stress builds up around it. When the stress becomes concentrated to a high enough degree, the locked fault ruptures with a huge energy release, causing an earthquake. Earthquake records made at seismic stations around the world show that this happens in all earthquakes. The different movements which resulted from adjacent blocks sliding past each other, moving apart from each other, or converging on each other, also occurred as expected according to the theory of plate tectonics.

Another study was made of the floor of both the Atlantic and Pacific Oceans. If the ocean floor has indeed been spreading out from oceanic ridges for some 200 million years, then during this time it would have been receiving a slow yet continuous rain of fine particles which would filter through the water and accumulate on the seabed. This sediment, made up mainly of minute marine organisms and fine clay eroded from the land or from volcanoes, should be thicker at the oldest parts if the ocean floor has indeed spread, and the age of the lowest layer should be almost the same as that of the crust on which it rests. Deep-sea drilling showed that the thickness of seabed sediment and age of the lowest layer gradually increased as measurements were taken farther away from oceanic ridges, as predicted by the theory.

Although most scientists now agree that the moving plates are fact, a question remains: what force is powerful enough to propel them along? How can continents float along on the dense, heavy liquid mantle? One suggestion is that the underside of the crust is being continually churned up by currents within the hot mantle. These currents, called *convection currents*, occur where hot, light material, either gas or liquid, rises naturally, and cooler, heavier material sinks to replace it. Convection currents are the basis behind the simple classroom demonstration of heating a mixture of oil and water: the hot oil rises through the less dense water, then gradually cools and falls to the bottom of the container, only to rise again when sufficiently heated.

The Dutch geophysicist F. A. Vening Meinesz originally put forward the theory of convection currents after he recorded unexpected changes in the earth's gravity during a submarine survey of the Pacific Ocean in 1923. Today most scientists are convinced of the existence of not one but five major churning convection currents within the mantle. It seems that the bedrock beneath our feet is in constant motion.

Was the earth once half its present size? Several scientists have recently begun to investigate this radical proposal. If it were true, then all the continents and other land masses would have covered the entire globe, leaving no room for the oceans. Has the earth gradually grown over the last 3000 to 4500 million years, rupturing its crust in places and slowly forming the ocean basins, finally reaching its present size? This may have occurred through a slow and inexorable weakening of the pull of gravity over vast eons of time. At the moment there is no device accurate enough to measure the earth's expansion—less than a sixteenth of an inch each year—if it is indeed taking place at the rate required by the theory. As yet it is impossible to disprove this theory, but future decades may reveal the solution to the mystery.

Below: early morning view of several Himalaya mountain peaks in Nepal—Taboche (foreground), Nuptse, Everest, Lhotse, Lhotse Shar, and Amaidablang. According to Alfred Wegener's theory of plate tectonics, the Himalayas were formed when India drifted northward and collided with the Asian continent after Gondwanaland split. The overriding of India by Asia in the collision zone crumpled, compressed, and folded the land into mountains.

Chapter 3
The Mystery of Evolution

Many classical Greek philosophers thought life was composed of four primary "elements"—earth, fire, water, and air. Religious creation myths claimed that life was created suddenly by an act of God. Both tended to assume that the various forms of life began exactly as they are now without the necessity for intermediate development. But in 1859 Charles Darwin published his monumental work *On the Origin of Species*, and with it he introduced not only a theory of evolution but a mass of observational evidence to support it. Darwin's formulation of natural selection has now emerged from the heated controversy which surrounded its first publication to become an accepted part of modern scientific thought.

"Both in space and in time we seem to be . . . near . . . to that mystery of mysteries, the first appearance of beings on earth." These words were written in his journal by a young, unknown, unpaid naturalist after he had visited the Galápagos archipelago, a group of oceanic islands thrusting straight up from the ocean floor some 600 miles west of Ecuador. His name was Charles Darwin, son of a wealthy doctor who had signed on as naturalist on the HMS *Beagle* for its survey voyage of South America and parts of Australasia. Of all the places Darwin visited, none impressed him more than this cluster of islands, "a little world in itself" inhabited by species of birds, lizards, and tortoises to be found nowhere else in the world. The four weeks he spent there marked a turning point in his thought and formed the basis of his ideas on the changing forms of animals and plants, which would astound the world some 24 years later when he published his concepts and theories of evolution.

Darwin was particularly fascinated by the island finches, which were similar to a type of finch he had seen on the South American mainland. However, on the Galápagos they came in 13 varieties instead of only one, and Darwin noticed that no two types of finch competed for the same kind of food. Later he wrote in his journal, "When I see these Islands in sight of each other and possessed of but a scanty stock of animals, tenanted by these birds but slightly different in structure and filling the same place in nature, I must suspect that they are only varieties [of an origi-

Opposite: colossal robber crab (*Birgus latro*) shinning up a coconut palm on a Fijian island. The adult crab's armor is not only stout enough to withstand a fall from a tree, but also helps to seal moisture inside the body, since drying out is one of the dangers facing a crustacean on land. On the continents healthy competition from a host of land-based insects and vertebrate predators has kept crabs firmly in the sea. But on the islands of the Indian and Pacific oceans these crustaceans have adapted to life on land, where they can exploit food resources that would normally have been commandeered by other animals. This so-called coconut crab is known for its habit of climbing up coconut palms, but whether it actually picks the fruit remains a point for debate. A fully grown robber crab measures a yard across and has muscular pincers that operate as both wrenches and hammers. A crab opens a coconut by first removing the outer husk fiber by fiber, then hammering the area around the "eyes" until it has made a hole large enough to admit a slender claw, with which it scoops out the meat. Sometimes a crab will even carry a coconut up a tree and drop it onto a stone to crack the shell.

Above: small tree finch (*Camarhynchus parvulus*), one of the finches whose differing beaks intrigued Darwin on his visit to the Galápagos islands. All the land animals found there came over the sea, and very few species have established themselves—just two kinds of mammals, five reptiles, six songbirds, and five other land birds. Variations from island to island can be numerous, but up to 10 different species can also be found on a single island.

Right: woodpecker tree finch (*Camarhynchus pallidus*), a small insect-eating bird. One of the most remarkable of Darwin's finches, it has evolved the beak but not the long tongue of a woodpecker, so it carries a twig or cactus spine with which it pokes out insects from crevices in the tree bark. This is one of the few instances of tool-using by an animal.

nal species].” He found particularly strange the range of beak sizes in the different species of *Geospizinae*, “from one as large as that of a hawfinch to that of a chaffinch and even to that of a warbler.” It seemed to him that from an original scarcity of birds in the archipelago, “one species had been taken and modified for different ends.”

The voyage of the *Beagle* took place between December 1831 and October 1836. A mass of observations and collections of animals, plants, and fossils, as well as later studies and experiments at home in England, combined to give Darwin the idea of evolution, which is no more than that all the varied kinds of animals and plants living on the earth today have developed as alterations of earlier types through natural changes over long periods of time. Darwin announced his theories in 1859 with the publication of his major work, *On the Origin of Species by Means of Natural Selection, or the Preservation of Favoured Races in the Struggle for Life*. The alternative title of the book described what was happening all over England in Victorian times—in the back streets, mills, and workhouses—and the general public was very interested in the idea of “favored races.” Perhaps this is one reason the *Origin* sold out its first print order of 1250 copies on the day of publication.

Although Darwin’s writing style is rather dull and tedious, and sometimes the reasoning seems muddled, his work provided undoubted proof that the different members of the plant and animal kingdoms of today are what they are partially as a response to environmental conditions, and that individual species have changed over time. This concept was one of the most important contributions ever made in the field of science. Until this time people had, on the whole, been quite happy to accept any of a variety of mysterious explanations of the presence of living creatures on earth. After all, the notion of evolving life does not come easily or immediately to the majority of minds. It was enough to know that an assortment of widely different living things existed—cats, dogs, bluebells, oak trees, horses, snails, and so forth. Few asked, “Where did they all come from?”

However, that does not mean philosophers and scientists were not thinking about the problem. In early times the origins of life were explained in symbolic form as myths, usually linked with the four primary “elements” of earth, fire, water, and air. The Greek philosopher Thales (640–546 B.C.) believed that all life began as water, while his contemporary Anaximenes of Miletus thought air was the prime source. Later some quite fantastic and rather creative ideas were put forward. For example, the 17th-century German Jesuit scholar Athanasius Kircher thought orchids gave birth to birds and small men. In the early 18th century the Frenchman B. de Maillet suggested that birds were derived from flying fishes, lions from sealions, and men from mermen. According to another popular doctrine, called *catastrophism*, the characteristic fauna of each period were destroyed by some type of gigantic catastrophe and replaced by a new animal population, either specially created or composed of creatures which migrate in from distant lands. The Flood of Noah’s time was believed to be the most recent catastrophe.

Many religions, of course, claimed that life was created by an

Secret of the Finches’ Beaks

Below: an early view of the origin of various forms of life. According to a botanical work of 1609, “There is a tree—not, it is true, common in France, but frequently observed in Scotland . . . From this tree leaves are falling; upon one side they strike the water and slowly turn into fishes, upon the other they strike the land and turn into birds.”

Aristotle's Form Theory

act of God. The Hebrews accepted the sequence of events as set out in the Old Testament. They believed that all living things were the result of a series of "special creations" performed in six days by an all-powerful personal God, although the six "days" were frequently interpreted as six epochs. The theories of many classical Greek philosophers also include some process of evolution or change. Aristotle, for example, showed an understanding of the importance of change: he maintained that a chain extended in series from the simplest forms of life, or "essences," through to the more complex. The most complex forms became progressively more perfect. He believed that this was brought about in

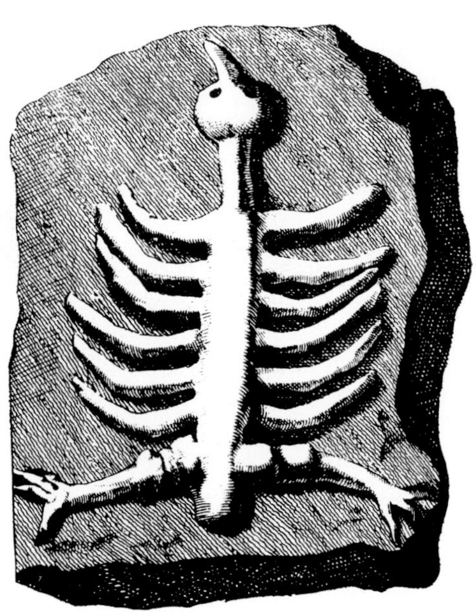

Above: one of a collection of "lying stones" which gave rise to a bizarre scientific hoax in the early 18th century. The victim was Johann Beringer, a German physician and professor at the University of Würzburg. Beringer was interested in the study of "formed stones," which today we call fossils. Two malicious and unscrupulous colleagues bribed three youths to take him some remarkable stones which bore on them in bold relief the forms of animals, plants, astronomical bodies, and even hieroglyphs resembling Hebrew letters. Thoroughly taken in by the stones, which were actually carved by one of the conspirators, Beringer published a book, *Lithographiae Wirceburgensis*, in 1726 in which he examined all the theories put forth to explain the amazing stones. Beringer is still remembered as the archetypal gullible scholar, so blinded by his passion for a new science that he failed to realize he was being deceived.

Right: the frontispiece to Johann Jakob Scheuchzer's catalog (including 528 items) of his Museum Diluvianum, published in 1716, in which he discovers traces of the Flood in the mountains of Switzerland.

the course of time by some mysterious perfecting mechanism. However, Aristotle's ideas were concerned with individuals, in contrast to later evolutionary thought—he thought in terms of one individual changing into a different individual rather than the gradual alteration of a species.

The religious claim of special creation and Aristotle's theory of "forms" went unquestioned for centuries, existing side by side throughout the Middle Ages, the Renaissance, and the Reformation. Until the late 18th century it was even accepted that the earth had been created in exactly 4004 B.C., a date calculated by the Archbishop of Armagh in the 17th century. However, as the science of geology advanced rapidly and fossils were discovered and studied, several scientists began to suggest that the world was several hundred thousand years old. Fossils had previously been thought to develop from moist seed-bearing vapors, blown from the seas into the crevices of the earth. Robert Hooke and Martin Lister, two 17th-century British scientists, were the first to suggest the possible relation of fossils to the geological strata in which they were found, although they did not necessarily believe that the fossils were the remains of decayed animal and plant life.

Then in 1795 James Hutton, an Edinburgh doctor, published an essay entitled "A Theory of the Earth." He claimed that divinely ordained floods and massive rock upheavals had created the oceans and mountain chains of the present world, and indeed that these forces were still at work. His theory, called *uniformitarianism*, also suggested that the continents had been formed over a period of millions of years, not merely a few thousand, and that the earth's future would be equally vast.

Yet, although by the beginning of the 19th century the concept of a changing earth was becoming accepted, still the idea that individual plant and animal species changed with time had not been suggested. Meanwhile, however, the early 19th-century British geologist William Smith was establishing the value of animal fossils in the recognition of different strata in geological formations. In France the naturalists Georges Cuvier and Jean Lamarck, Smith's contemporaries, were providing the foundations of invertebrate and vertebrate fossil history. These great spurts forward in the geological sciences provided a wealth of information about plant and animal characteristics, including the documentation of many extinct plants and animals. Gradually the idea of a succession of animal types through time began to unfold, although no one actually crossed the threshold of an evolutionary interpretation of life.

In the 1830s exciting discoveries were made by the naturalist Karl Ernst von Baer in Germany, who showed that the early embryo of a dog is indistinguishable from the embryo of any other mammal, and indeed from that of a fish, bird, amphibian, or reptile at the same stage. Later it was shown that even the human embryo passes through stages similar to those of the lower mammals and vertebrates in the course of its development. But even with this evidence, no theory of evolutionary change was forthcoming at this time, although it was staring scientists in the face.

Jean Lamarck was the first, in the opening years of the 19th century, to attempt to solve the problem of how evolution had

Above: Georges Cuvier (1769–1832) working with fossil bones, 1823. In order to explain the existence of different fossils in each rock layer, Cuvier invented his catastrophic theory. He suggested that floods and earthquakes periodically devastated parts of the earth and destroyed living things in those areas, and that new forms of life from other areas later repopulated them.

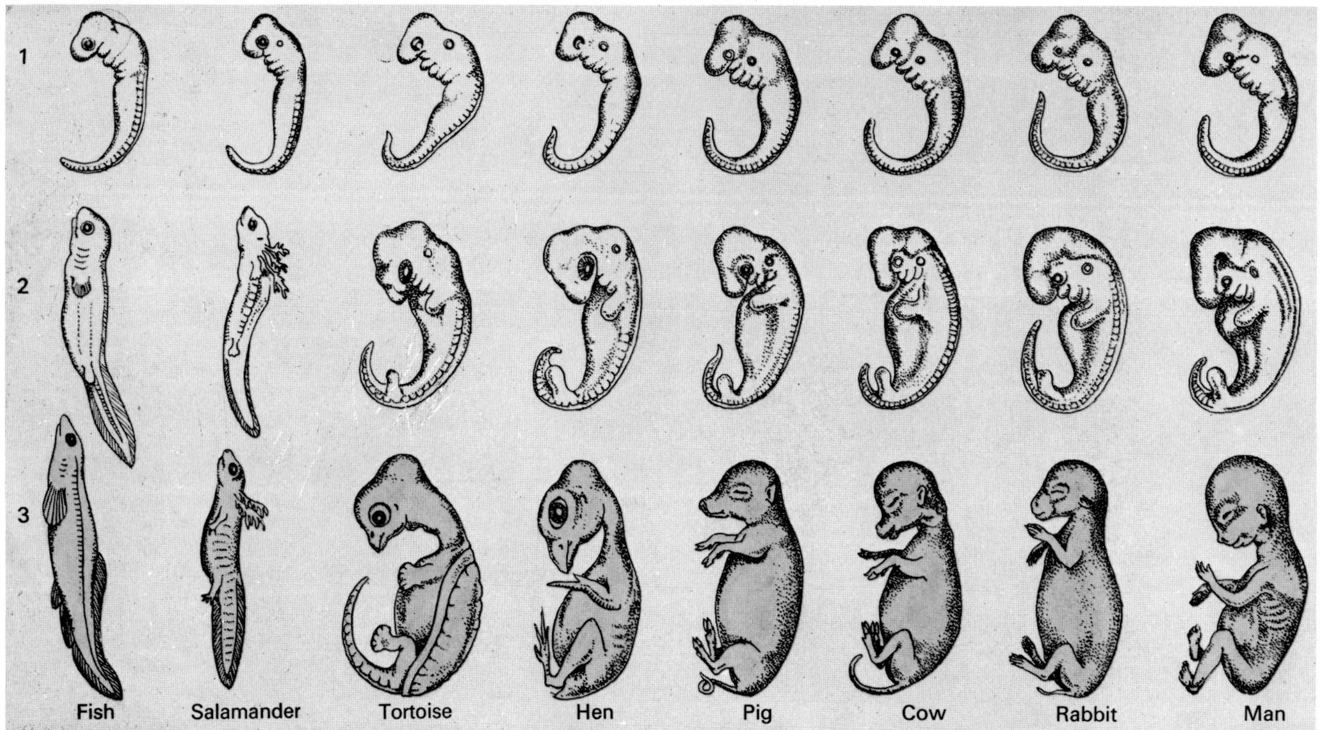

Above: stages in the early growth of eight animals built on the vertebrate plan, showing their similarity at the earliest stages. They are all descended from one original ancestor.

occurred. He suggested that species might have been transformed to other species through the "inheritance of acquired characteristics." Boldly he proclaimed that no essential difference existed between species and their individual varieties, that both were subject to change, and that "transformation," not immutability, was the basic quality of life. In *Zoological Philosophy*, published in 1809, he pointed out that living things change in form during their lifetimes depending on what they do—for instance, a man can build up his muscles by repeated exercise. Lamarck proposed that such "acquired characteristics" could be inherited by the children of the next generation, who would be born with slightly stronger muscles. He argued that if these changes were to continue indefinitely, the result would be a new species.

The classic Lamarckian example explains how the giraffe obtained its extraordinarily long neck. He suggested that this species could have evolved from some ancestral short-necked type, which generation after generation strained to reach ever more highly-placed leaves on trees, and that each generation inherited the slightly longer neck developed during the lifetime of its parent. This was the basis of Lamarck's theory—that species gradually changed in form by the use or disuse of a particular organ or faculty. Although Lamarck was wrong about the way evolution took place, he was certainly on the right track when he introduced the idea that species are not immutable. However, he could only speculate, since in the course of a considerable number of experiments he failed to produce evidence that any creature could pass on acquired characteristics to the next generation.

Lamarck's theory of evolution was later emphatically crushed by the German biologist August Weismann, who between 1868 and 1876 published a series of papers on the inheritance of variation. He firmly believed that acquired characteristics could not be inherited. He made little use of practical evidence, although he did carry out a series of crude experiments with many generations of mice, which showed that the offspring were always born with tails whether the parent's tail had been cut off or not. His hypothetical reasoning assumed that there was a distinction between the body of an organism—the *soma*—and the germ cells concerned only with reproduction—the *germ plasm*—which led to the formulation of his "theory of the continuity of germ plasm." He stipulated that only the germ plasm could affect inheritance and that the soma played no part in it. He also identified *chromosomes*, the thin bands which are present in every germ cell and which carry the genes, and then went so far as to develop a theory of inheritance based on chromosome behavior. Weismann's work was a serious blow to Lamarck's theory, which was largely discredited by the 1880s.

However, in the interval between publication of Lamarck's theory and Weismann's rebuttal, Charles Darwin had attracted the startled attention of the Victorian public and the scientific world with his painstakingly documented theories of evolution. *On the Origin of Species* quickly became the center of a stormy debate, and Darwin himself was acclaimed as a genius. Considering his unexceptional early childhood and mediocre school and university record, Darwin's later brilliance comes as something of a surprise. Born in Shrewsbury, Shropshire, in western England, in 1809, son of a wealthy doctor, Charles received the usual education in Latin and Greek verse, geography, and history at the local private school, and he later remarked when writing his autobiography that school to him "was simply a blank." However, he read extensively in his spare time, delighting in the works of Shakespeare and Lord Byron; *Wonders of the World* filled him with a desire to travel. *Selbourne*, by the 18th-century English naturalist Gilbert White, helped to deepen his interest in nature, especially ornithology. Even at the age of 10, he was interested in insects, chiefly butterflies and moths, and collected them near his home and while on vacation at the Welsh coast. He noted that many specimens in his collection from Wales were not to be found at home in Shropshire.

Weismann-or Lamarck?

Above: Charles Darwin (1809–1882), who conceived one of the most important theories in modern science.

Below: garden view of Down House in Downe, Kent at the time Darwin lived there. He and his wife Emma moved there with their two oldest children in 1842. The seclusion he found here, in addition to the freedom from financial worry insured by the generosity of his father, enabled him to work out his theories of evolution over a long period of time.

Charles Darwin Joins the Survey

His father wanted Charles to follow in his own footsteps by studying medicine at Edinburgh University in Scotland. Darwin went, but he was revolted by his first two operations—not surprisingly, since anesthetics were not yet in use. After two years it was quite obvious that the 19-year-old Darwin was unsuited for his father's profession, and so he returned home. Charles did not voice any objection when it was decided that he should go to Cambridge University to study for the church instead, and indeed he entered Christ's College. However, he frequently neglected his studies in favor of enjoying "wine, women, and song" to the full, although he also spent hours shooting, collecting beetles (he even hired a helper, whom he sacked for giving the best specimens to a rival), and he spent a great deal of time with Professor J. S. Henslow, a professor of botany. However, Darwin did manage to pass the exams for his degree, and he was resigned to taking church orders.

At this time there was no question of his doubting the strict and literal truth of every word of the Bible. However, fate intervened and set this apparently unambitious man, unremarkable in either studies or name, on a path which would show the world his hitherto unseen qualities of painstaking observation and sound reasoning. On the advice of his friend Professor Henslow, head of the botany department at Cambridge (and with the support of his beloved Uncle Jos—Josiah Wedgwood—who persuaded his unwilling father to let him travel and give him money), Darwin joined the survey ship HMS *Beagle* under Captain Robert Fitzroy as an unpaid naturalist on a journey to South America and Australasia.

When the *Beagle* finally set sail on December 27, 1831 Darwin suffered immediately from seasickness (which would afflict him throughout most of the voyage). Darwin had studied the British geologist Sir Charles Lyell's *Principles of Geology* and had acquired a general background in "uniformitarianism" rather than "catastrophism." During the following months Darwin's observations forced him to doubt the entire doctrine of immutability of species, suggesting instead that species evolve. When traveling through South America, he observed that a species occupying a particular niche in nature in one region was replaced in nearby areas by other species that were definitely different and yet quite similar. He questioned why the rabbitlike animals of the La Plata savannahs were a type of rodent peculiar to South America, instead of being like those of North America or the Old World. He also wondered why the species of rhea which lived on the Patagonian plains was replaced on the pampas by another species of South American rhea and not by an Old World ostrich. On the pampas Darwin found fossil remains of huge mammals who had been covered with bony armor like that of the various armadillos currently living in the same region. He wanted to know why these extinct animals were physically similar to the living ones.

By the time Darwin returned to Britain on October 2, 1836, he was convinced that a process of evolution occurred, but how the necessary changes had taken place still eluded him. He had various ideas and wrote numerous notes, but he kept them from all but a few close friends.

Opposite: case of the beetles Darwin had collected from an early age, preserved at Down House in Kent.

Below: the H.M.S. *Beagle* lying at anchor in the harbor at Sydney, Australia. The *Beagle* was a very small brig, only 90 feet long, but at the time of Darwin's memorable voyage it carried over 70 people. Under the command of Captain Fitzroy, it had been commissioned by the British government to survey the southern coasts of the Americas, southern islands, and Australasia. Darwin's experience of the entire five-year voyage was marred by constant bouts of seasickness.

Above: the nine-banded armadillo (*Dasypus novemcinctus*), best-known and most widely distributed of the armadillos. Darwin compared this living animal to the fossils he found in South America of the extinct Glyptodon (top), which had a bony armor covering its skeleton.

Over the 10 years it took him to write up his findings from the *Beagle* voyages, Darwin led a quiet and secluded life at Downe, in Kent, England, having married Emma, youngest daughter of his favorite Uncle Jos. Eventually his work began to cover ever-widening horizons. He was interested in everything and wanted to try out many ideas. He studied all kinds of domestic animals: sheep, cattle, pigs, dogs, cats, poultry, goldfish, and pigeons. He also studied canaries, peacocks, bees, silkworms, earthworms, flowers, and vegetables. Darwin, a devoted family man who loved all of his numerous children dearly, found that because of their demands on his time his work progressed only gradually. Slowly Darwin pieced his theory together, gathering so many observations on variations in domestic breeds that finally he could not doubt that the accumulation of favorable variations over long periods of time must result in the emergence of new species and the extinction of older ones.

To the Victorians, this was heresy. The average Christian believed every word in the Bible, including the Book of Genesis, and here was Darwin seeming to deny the role of God himself. Partially for this reason, Darwin kept his views to himself for almost 20 years, except for his close scientific friends Sir Charles Lyell and Joseph Hooker, the great botanist. Darwin's beliefs would probably have died with him had it not been for Alfred Russel Wallace, another British naturalist, who wrote to him from Malaya asking him to read an essay entitled "On the Tendencies of Varieties to Depart Indefinitely from the Original Type." He read it and was astounded that another man was thinking along the same lines as himself. Wallace had asked Darwin to pass the essay on to Lyell if he thought it good enough, and Darwin showed no hesitation in doing so, although after warmly recommending the essay he did add a note to Lyell: "So all my originality will be smashed." Fortunately Lyell and Hooker insisted that he publish a joint paper with Wallace, and a month later this was presented to the Linnaean Society, in July 1858. Darwin got to work and, with Wallace's blessing, published *On the Origin of Species* the following year.

Darwin explained evolution by the principle of natural selection. Put simply, this means that every species of living thing produces more offspring than can possibly survive on the available amount of food. Only those individuals equipped to deal successfully with their environment in such a way as to "win out" at the expense of their competitors will survive to reproduce themselves, resulting in the long-term survival of the "fittest" of each species. Darwin visualized the struggle at two levels. Sometimes only the weak individuals of a single group will perish. But if all the individuals are at a disadvantage when compared to other competing groups, the entire species may become extinct. This is what is thought to have happened to the dinosaurs, although the exact reason for their downfall is still a mystery.

Darwin also noted that no two living things are born exactly alike, with the possible exception of identical multiple births. Though often small, the differences between individuals are occasionally important. For example, an animal with a coloring which camouflages and hides it from predators in its environment has an advantage over other members of its kind which are

Left: anti-Darwinist cartoon of 1882 from *Punch*, titled "Man is But a Worm." It links the worm, emerging from primeval ooze, with man and a sympathetic, even affectionate, representation of Darwin himself. This was just one of many hostile cartoons popular after the publication of Darwin's theory of evolution.

The Influence of Wallace

not perfectly camouflaged, such as an albino. Holders of these advantageous variations will therefore be more likely to succeed than others in the competition for survival. Parents of the next generation will tend to come from those members of the species best able to cope with the conditions of their environment. Subsequent generations will maintain and through gradual change improve upon the advantageous characteristics of the parents.

It seems amazing that Wallace hit upon the same solution for the mutability of species. And yet is it? Wallace's ideas were based on observations made in South America and the Malay archipelago, and his approach was largely geographical. He was struck by the fact that *orders*, large animal classification groups, are usually distributed widely over the world. For example, the marsupials or pouched mammals survive in both Australia and South America, and fossils have been found in Europe and North America. Smaller groups, such as families and genera, have only a localized distribution, as with the kangaroos of Australia. He also noted, "When a group is confined to one district, and is rich in species, it is almost invariably the case that the most closely related allied species are found in the same locality or in closely adjoining localities, and that therefore the natural sequence of the species by affinity is also geographical." Kangaroos again are a good example, since they are confined to the Australasian region, and all the species of kangaroo known today resulted naturally by evolving from a common ancestor living in the same region millions of years ago. Wallace showed that the fossil

Below: Alfred Russel Wallace (1823–1913). A much-traveled animal geographer, he shared with Darwin the rare ability to combine original observations with deduction. He arrived independently at much the same conclusions as Darwin.

record was not mysterious but showed a simple succession of forms. Although much of the credit for establishing the fact of evolution is rightly credited to Darwin, it is only fitting that Wallace should also be recognized and honored as an original and imaginative thinker.

Most of the bitter attacks which followed the public unveiling of the evolution theory missed Wallace, since he was still living in Malaya. Due to his chronic ill-health, Darwin also managed to stay out of much of the public debate, leaving Thomas Henry Huxley, the most brilliant zoologist in England at the time, Hooker, Lyell, and others to carry on the battle. Darwin's most dangerous scientific opponent was Richard Owen. He was an outstanding comparative paleontologist and anatomist, and he briefed Bishop Wilberforce, nicknamed "Soapy Sam," for the famous debate on evolution which took place at the Oxford meeting of the British Association in 1869. Unfortunately for Owen and the bishop, T. H. Huxley, though only 35 years old, was in brilliant form, and he routed the bishop and his followers. Huxley continued successfully to repel anti-Darwin attacks and christened himself "Darwin's bulldog."

One problem was that, at the time Darwin worked, experimented, and wrote, the principles of heredity—how characteristics were transmitted from one generation to another—and of how mutations occurred were completely obscure. The missing piece was being discovered, however, at this very time, unknown to Darwin or anyone else. Careful experiments were underway in the monastery garden of Gregor Johann Mendel.

Below: brush-tail opossum (*Trichosurus vulpecula*), the most common and widely distributed of all Australian marsupial species. It is often found in gardens on the outskirts of cities in Australia, Tasmania, New Guinea, and New Zealand. About 2 feet long, arboreal, and very active, it has a foxlike head with large ears and a pointed snout. It uses its bushy tail, with the tip of which it can hold onto branches, to balance.

Below right: murine or mouse opossum (*Marmosa murina*), common in Central and northern South America. The females have no pouch at all but carry their young on their backs. Alfred Russel Wallace noticed that animal orders such as the marsupials are usually distributed over several continents, while closely related families of species are found in the same locality.

Gregor Mendel and Heredity

Mendel was born to peasant parents on July 22, 1822 in the small Moravian village of Heinzendorf, now in Czechoslovakia but then part of the Austro-Hungarian empire. As a child he was luckier than most peasant children in that his father was the first member of his family to own his own farm, although since he still worked three days a week for the lord who had originally owned the land, life was not easy, and there was little money. Mendel learned to graft trees in the orchard, and his farming childhood may have inspired his later work. Mendel was also fortunate in that his parents managed to send him not to the village school but to a better one some miles away. The boy did extremely well, although he had to survive mostly on bread and butter to help pay for his schooling.

Mendel had hoped to go on to a university, but his hopes were dashed when his father fell ill. He was only 20 when he decided to enter the monastery at nearby Brunn and become a monk of the Augustinian order. Five years later, in 1847, he was ordained a priest, and was called to be a temporary teacher at a nearby school. He was promised a permanent position if he passed some government tests, but failed them—though this was not to prove an obstacle since he so impressed the examiners that he was sent to the university in Vienna. Two years later he returned to teach science at the local high school, and he was able to grow plants and to experiment with breeding in the monastery garden. He began exploring the mysteries of heredity.

Mendel's experiments on ordinary garden peas were remarkably simple. Mendel noticed that not all the peas in his garden were alike: some were tall plants while others were short; some had yellow and others green peas; some had white flowers while others had red. Mendel sorted out some 20 differences and carried out a program of experimental breeding. The process of crossing two dissimilar strains is called *hybridization*, and the offspring are called hybrids. Mendel bred tall plants with short, green pea seeds with yellow, and so on. The pea is particularly useful for controlled hybridization experiment because it is usually self-pollinating, and cross-fertilization occurs only when some external agency (in Mendel's case a fine paintbrush) transfers the pollen from one plant to another.

Mendel called his plants "my children" and obviously loved his work. He found that when tall pea plants were crossed with short, they produced only tall offspring. When these tall hybrid offspring were mated with each other, about 75 percent of the next generation were tall and 25 percent were short. Because tallness "covered up" shortness characteristics in the hybrid peas, Mendel called tallness a *dominant* trait, and the shortness a *recessive* trait. He found from other experiments that a yellow seed coat color was dominant over green, and that red flowers were dominant over white.

In order to explain the constant ratio of three-to-one, Mendel suggested that the seeds contained factors or genetic elements—today they are called *genes*—which could be regarded as "chemical messengers" which pass on information from parents to offspring. The appearance of each trait was due to the influence of a particular genetic element. Mendel worked with his peas for eight years and carried out over 10,000 experiments. In the course

Below: Gregor Johann Mendel (1822–1884), scientist-monk who laid the foundations for the modern theory of genetic inheritance. He showed that characteristics did not blend but were transmitted as either dominant or recessive traits within sex cells.

The Importance of Mutations

of his work with first, second, and third generation hybrids Mendel discovered some of the fundamental laws of heredity. He found that each trait is controlled by two factors (genes), one inherited from each parent. Each of these genes is either dominant or recessive. For every trait, a new pea plant could inherit either two dominant genes, two recessive genes, or one dominant and one recessive gene. His last but extremely important discovery was that the distribution of dominant or recessive genes from parents to offspring is determined by chance.

Today it is accepted by biologists that Mendel's work established the basic laws of inheritance and provided a clear and factual basis for the science of genetics. Why did his work go unnoticed, and the laws of heredity remain a mystery to the scientific world for 40 years? Mendel was unfortunate in that initially in 1865 he read his paper only to the Brunn Society for the Study of Natural Sciences. The 40 or so citizen-scientists were so little interested that they did not even ask questions, no doubt deeply involved in their own particular areas of research and unable to grasp Mendel's array of statistics. Although some scientists present were working with the new Darwinian ideas on evolution, it was difficult for them to see how Mendel's statistics could provide a mechanism whereby small combinations of characteristics could be reshuffled within a preexisting species. Mendel did not actually suggest that his laws were a method through which genuinely new species could be produced, although he had read and studied Darwin's *Origin*.

Unlike Darwin and Wallace, who were encouraged by scientific friends, Mendel was rebuffed. He submitted his paper to Karl Nägeli of the University of Munich, a leading botanist of the day. Although interested in the current evolutionary con-

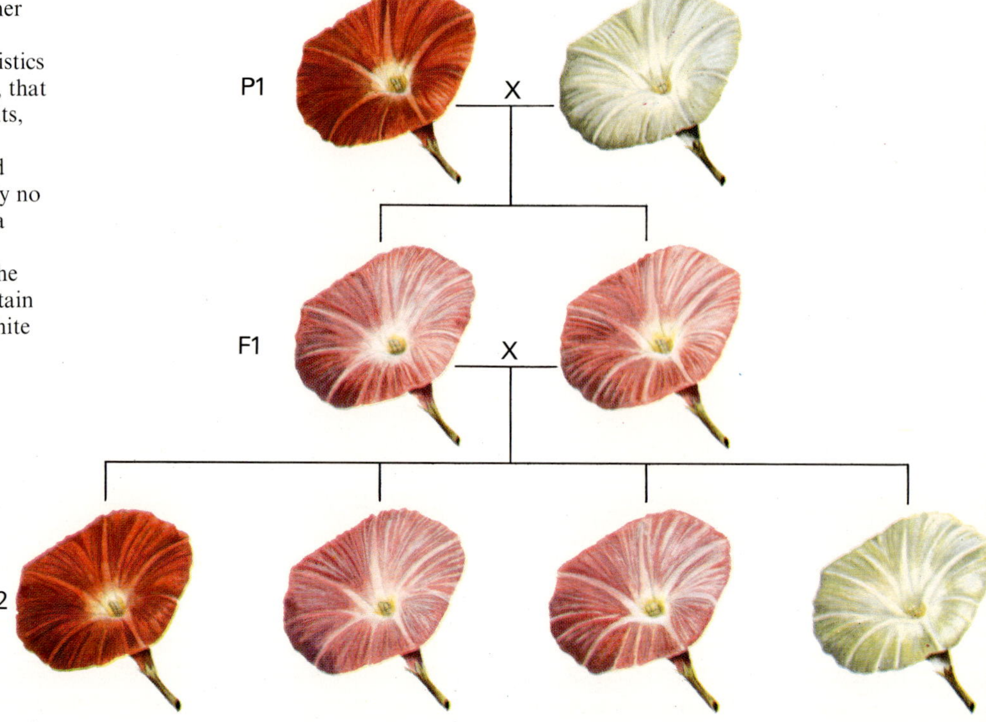

Right: Mendel dealt primarily with genes in the pea plant, which clearly showed either complete dominance or complete recessiveness; the blending of characteristics did not appear. Mendel knew, however, that some plants *appear* to have blended traits, as is the case with the morning glories shown here. Even so, the genes involved still obey Mendel's laws—there is simply no dominant gene for color, so that when a gene for red and a gene for white occur together the result is a pink flower. In the third generation we would expect to obtain a ratio of one red to two pink to one white flower (1:2:1).

troversies, Mendel's paper had little effect on him, and he sent back a rather brusque letter in reply. "You would do better," said Nägeli, "to work with another plant." Mendel, assuming the famous botanist knew more than he did, followed his advice and began to experiment with other plants. But in choosing the pea Mendel had been lucky, as it has a fairly simple genetic mechanism. The new plants were not so simple, and none seemed to follow his early findings. Discouraged, disillusioned, and broken-hearted, Mendel spent less time on his "hobby," and on appointment as abbot of his monastery he abandoned his research altogether. He died in 1884, his papers on breeding having gathered dust for 20 years, although they had at least been printed in a scientific journal where some future interested person could find them.

It was not until 1900 that Mendel's work was independently rediscovered by four different biologists: Hugo de Vries in Holland, William Bateson in England, Karl Correns in Germany, and Erich Tschermak in Austria. It was Karl Correns who summarized Mendel's conclusions in the form of two principles or laws of dominance, which have since been shown to be valid for all types of organisms. De Vries, meanwhile, had been studying evening primrose plants (*Oenothera biennis*), which had been introduced to Holland from America. In his cultured plants, de Vries discovered that a few completely new forms had appeared among the numerous ordinary forms, although they had all been growing under exactly the same conditions. The variants were quite distinctive, and each differed from the others. De Vries gave the name *mutation* to this production of new forms through sudden change. It is quite different from Darwin's gradual process of accumulation, which relied on continuous variation under the influence of natural selection. Mutation was a totally new concept of the origin of variation, and it accounted for the appearance of sharp, sometimes drastic, discontinuous variation in plants and animals.

Above: evening primrose plants (*Oenothera biennis*) in Hugo de Vries' experimental garden, covered with wire mesh. Over a 10-year period he raised 53,509 plants in an effort to understand the occurrence of what we know now as mutations. De Vries was suspicious of the Darwinian view that species are only modified gradually as a result of the piling up of many slight variations, and he had thus begun to look for evidence of sudden jumps ("saltations," T. H. Huxley had speculatively called them). The irony is that, as was later proved by a Dane who repeated his evening primrose experiments, what de Vries had classified as separate new species of the plant were actually only stable intermediate types of the same species.

Below: Hugo de Vries (1848–1935), the Dutch botanist who was one of the first scientists to rediscover the importance of Mendel's publications. He is also known for developing the experimental method of studying evolution through the observation of mutations rather than as the gradual result of natural selection.

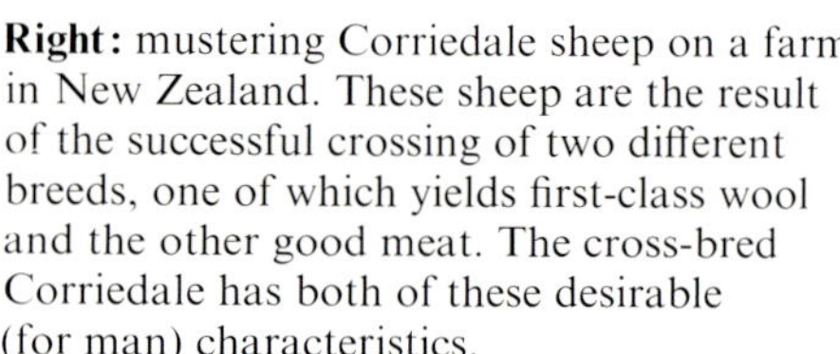

Right: mustering Corriedale sheep on a farm in New Zealand. These sheep are the result of the successful crossing of two different breeds, one of which yields first-class wool and the other good meat. The cross-bred Corriedale has both of these desirable (for man) characteristics.

De Vries published his observations between 1900 and 1903 as a "mutation theory," and he suggested that evolution found its basic raw material in sudden, spontaneous changes. Using Mendel's rediscovered work on inheritance, William Bateson and the American geneticist Thomas Hunt Morgan made it possible to replace speculation with sound experimental evidence. The more extreme believers among the scientific community abandoned all Darwinian ideas when continuous variation was shown to be insufficient to cover all phenomena, and they assumed that all new forms arose only as a result of some abrupt change in gene composition. The mutation theory took over and Darwin's fundamental principle of natural selection was cast aside. However, over the next 30 years or so, study in the field of genetics exploded, mathematical methods were adapted to test evolutionary ideas, and biological research revealed the answers to many mysteries concerning variation and inheritance. The modern concept of evolution requires Darwin's original concept of natural selection in addition to taking note of occurrences of discontinuous variation through abrupt mutation.

Although a tremendous amount is now known about evolutionary change, few biologists would say that we understand more than a fraction of the mysterious course of organic evolution. Explanations can be found in terms of variation, heredity, selection, and adaptation, although it is still difficult to grasp the importance of the added dimension given by an infinite time span. But during the last 60 years researchers have been able to demonstrate evolutionary changes in populations over time periods as short as one generation. One example is the adaptation shown by bacteria in the growth of their resistance to antibodies. Another instance is the divergence of populations to produce genetically isolated groups which exhibit adaptations for some particular mode of life, illustrated by Darwin's finches of the Galápagos.

Another level of evolutionary change is that of *adaptive radiation.* This occurs when a general adaptation turns out to be so effective that it is exploited rapidly over a wide variety of habitats, manifesting itself differently in each case. Once an ancestral species with the new trait has emerged, the group deploys into

What Exactly is "Adaptive Radiation"?

many forms, each displaying the same basic plan and reactions. However, each form becomes anatomically adapted to its own environment. A good example of this phenomenon is the trilobite group of invertebrates, which were characterized by their flattened, oval bodies. Now extinct, they flourished during the Jurassic period, around 150 million years ago. Some trilobites, for example, were minute in size and probably crawled over the sand and mud of the earth's seashores. Others were larger, about 18 inches long, with pronounced head shields which were useful for rooting in the seabed for food. Still others were extremely active swimmers, with lighter head shields and limbs which propelled them through the water at the surface. Other trilobites evolved adaptations suited for a floating life, with broad heads and tail shields and projecting spines which allowed greater buoyancy. So successful was the original adaptation that, over a relatively short time, a great burst of evolutionary change helped to distribute it through a large and varied population.

Another rare but important change is known as *mega-evolution*, in which important adaptations have meant an all-around biological improvement as opposed to specialized changes of a limited nature. The British biologist Sir Julian Huxley has stressed this type of evolutionary change, explaining that each new "characteristic" has given organisms greater control over, as well as greater independence of, their environment. Huxley pointed out various "steps" which have helped dominant groups to progress throughout the course of evolution. Among early primeval organisms, development of the organism's ability to manufacture its own food using sunlight, carbon dioxide, and water—known as *photosynthesis*—was an absolutely fundamental step. Without this the whole balance of nature, with its cycles of matter and energy, could never have developed.

Another step in mega-evolution for plants and animals was from a single- to multi-celled condition. Becoming adapted for certain types of movement, survival on land or in the air, as well as the development in mammals of warm-bloodedness, increased brain size, and ability to give birth to live babies are other major evolutionary steps. However, although there is now some agreement among scientists concerning evolution, this does not mean that the whole mystery has been solved, or even that it ever will be. The ultimate mystery of life may be beyond the reach of both scientific investigation and, indeed, the human mind.

Below: the wide span of the gannet's wings makes it perfectly adapted to take advantage of the slightest updraft around the cliffs and coastal waters it inhabits. About 100 million years ago the problem of drag was first solved by the evolution of an efficient wing. Since then, adaptive radiation has meant the widespread adoption among most bird species of this basic framework, while each individual species has developed a modified wing shape most suited to its lifestyle.

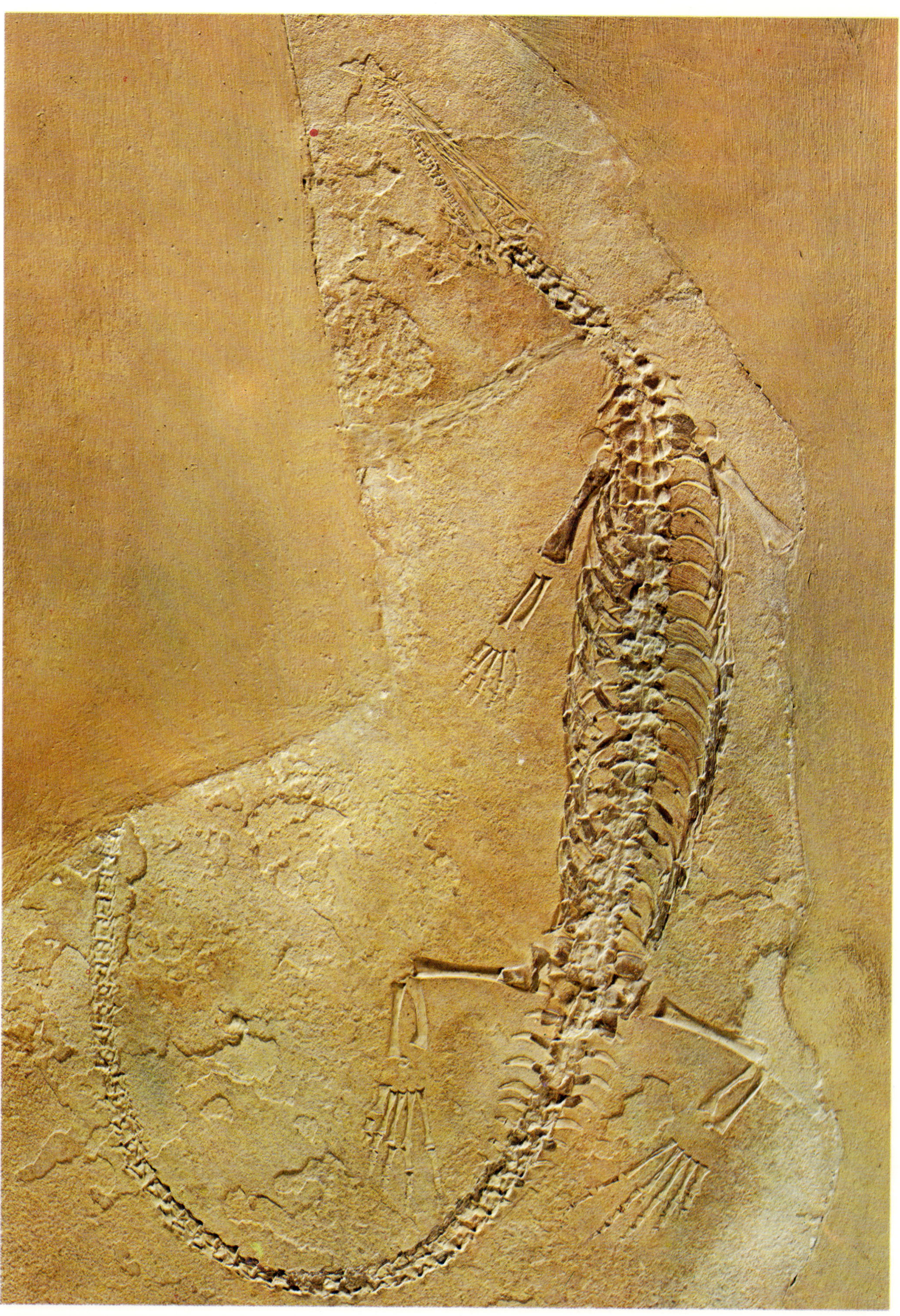

Chapter 4
The Riddle of the Dinosaurs

From a country lane in Surrey, England to the Gobi Desert, fossil teeth and eggs of long-extinct dinosaurs have been turning up in modern times to perplex and fascinate scientists. Interest had built up to such an extent by the 1880s that "dinosaur wars" were fought between rival groups of collectors. But the question which has generated the most speculation and controversy relates not to the *lives* of such gigantic reptiles as *Brachiosaurus*, *Diplodocus*, and *Scelidosaurus*, but to their deaths—why did the dinosaurs become extinct? Will we ever know the real answer to the problem of how the entire dinosaur population died off over a period of only 5 million years?

In early spring of 1822 Dr. Gideon Mantell, a physician, was on his way to visit a patient in the Cuckfield district of Sussex in southern England, accompanied in his horse-drawn carriage by his wife. As he made his call, his wife wandered down the country lane taking an interest in the surroundings. She stopped at a pile of broken stones, known as road metal in Britain because it was used to fill ruts and mend roads. The stones were of hard, calcareous or chalky grit known as Tilgate grit, and on some of them she saw large fossil teeth embedded. Entirely by chance, she had made the first discovery of remains of the dinosaur *Iguanodon*, although some time was to pass before they were correctly identified. She brought her find to her husband who was already fanatically interested in fossils, and indeed the teeth would in future become more important to him than even medicine or his family. Mantell knew immediately that in his hands were the remains of an animal new to science, but he was very puzzled as to what kind of animal it was. The teeth were worn smooth and each had a sloping (oblique) crown. This characteristic is typical of an herbivorous, or plant-eating, animal, but other mysteries still surrounded the teeth.

Mantell knew all the quarries in the district, and with the help of local quarrymen he discovered more teeth and bones and identified the geological layer in which they had been embedded. The layer had never been known to hold mammalian fossils, which led Mantell to believe that the teeth could not have be-

Opposite: fossil skeleton of *stereosternum*, an early reptile of southern Africa in the Permian period. It was one of the mesosaurs, a group of slender, aquatic fish eaters about 3 feet long. They probably derived from reptiles known as cotylosaurs.

Right: the Sussex quarry near Cuckfield where Dr. Gideon Mantell discovered the fossil dinosaur teeth. In the upper lefthand corner the spire of Cuckfield Parish Church can be clearly seen so that the location of the find is fixed. Mantell's diary records numerous visits to the quarry to search for more fossils. He published *The Fossils of the South Downs; or Illustrated Geology of Sussex* in May 1822, a few months after his wife had found the first teeth.

longed to a mammal. However, he did not believe that there was a reptile living at that time (such as a tortoise, crocodile, or lizard) that had teeth similar to his find, which had been designed for grinding and chewing food, so Mantell did not feel confident suggesting that he had the teeth of a reptile from ages past.

Mantell asked scientific friends and associates for help with the process of identification, and he received a wide variety of replies. The French naturalist Georges (Baron) Cuvier thought the teeth were the upper incisors of an extinct rhinoceros, while members of the Geological Society and others in London claimed that they were from a large fossil fish known as the wolffish, *Anarrhichas lupus*. An alternative suggestion was that the teeth belonged to an herbivorous mammal whose body had somehow become buried in an older geological strata.

Mantell was not easily discouraged by these eminent authorities and was still convinced that he had discovered the teeth and bones of a hitherto unknown herbivorous reptile. He continued his detective work in an attempt to prove beyond all doubt that the teeth were from an ancient plant-eating reptile. He searched every specimen tray in the Hunterian Museum at the Royal College of Surgeons in London, which in those days housed thousands of animal skeletons great and small. As in every good mystery story, a chance meeting solved the identity of the teeth. A young man, Samuel Stutchbury, was shown the teeth by Mantell and William Clift, the curator of the museum who was helping with the search. At once Stutchbury saw a similarity to the teeth of the living iguanas of Central America, reptiles he had studied in detail. Mantell was very excited at this, realizing that here was his missing clue. His fossil teeth were gigantic iguana teeth belonging to some prehistoric reptile.

In 1825 Mantell's discoveries were published. Even Baron Cuvier admitted his error and acknowledged that the paleontologists had a new subject to consider—a group of huge reptiles from a past age of a type not previously known, living lives completely unlike modern living reptiles. At this time the word "dinosaur" was still a generation away.

Mantell's *Iguanodon*, however, was not the first such discovery to be made. A very early find by a Sir Thomas Pennyston was recorded in 1677 and described by Dr. Robert Plot, the first

Below: earliest known drawing of a megatherium skeleton, made by a Spanish artist in the late 18th century. The skeleton was found in 1789 and the animal, the South American giant sloth, was named by Baron Cuvier in 1796. Cuvier (1769–1832), an extremely influential natural scientist in his own day, made a fateful and inaccurate pronouncement when he said that there was no such thing as fossil man.

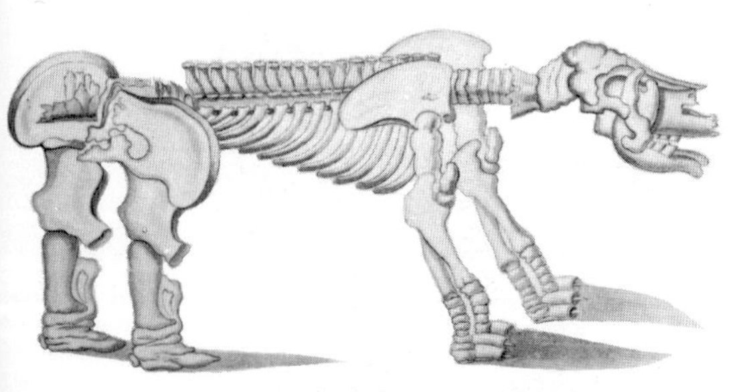

keeper of the Ashmolean Museum and professor of chemistry at Oxford University. It weighed 20 pounds and was identified as part of a thigh bone belonging to some huge animal such as an ox, horse, or perhaps an elephant. Unfortunately both the bone and further references to Sir Thomas have been lost without trace. A large "thigh bone" was found on the North American continent in 1787 by Dr. Caspar Wistar, an anatomist and one of North America's earliest scientists, and A. Timothy Matlack. It may well have been from a huge duck-billed dinosaur, but no scientific description was made.

Many other dinosaur bones were discovered in Britain and America toward the end of the 18th and early into the 19th century, but at the time their significance was not recognized, and not all were preserved for later study.

While Mantell was discovering fossil bones, Dean William Buckland, a clergyman and professor at Oxford University, was working on descriptions as well as discoveries. Buckland was of an inquiring, scientific turn of mind, and he devoted himself to the study of the geology of Britain, a field at this time in the formative stages of its development. Buckland must have been a delightfully entertaining man to know. His home housed not only his family but, from time to time, an assortment of strange—but living—animals. One amusing account, by a tutor to Buckland's son, tells of him coaching the boy one day when he noticed a jackal wandering around the room. The animal took refuge under the sofa from which spot then emerged a sound of munching and crunching. Later it was discovered that four or five guinea pigs had been the distraction and had thereby met their

Dr. Mantell's Iguanodon

Below: dinner in the *Iguanodon* model at Crystal Palace in Sydenham, south London, in 1853. It was written afterward, "In the mold of this colossal work of art . . . the idea was conceived of bringing together those great names whose high position in the science of paleontology and geology would form the best guarantee for the severe truthfulness of the work . . . and in the head of the gigantic animal sat Professor Owen . . ."

Above: Edward Drinker Cope wrote his first notes on fossils at the age of six and at 19 published a paper on salamanders. He was determined to better his loathed archrival, Othniel Charles Marsh.

Below: Othniel Charles Marsh (center, back row) and his students dressed for a dig. Marsh could not afford an education until he was 21, but he went on to become a co-founder of the Peabody Museum of Natural History at Yale University.

deaths. Another well-known member of the Buckland menagerie was a bear, which attended wine parties dressed in a scholar's cap and gown, much to the amusement and horror of the guests.

However, Buckland still managed to accomplish much respected work on fossil reptiles. In 1824 he described strange bones and a jaw 40 feet long that had been found near Oxford, and he decided that they belonged to a gigantic, elephantine, meat-eating reptile. He named it *Megalosaurus*, from *megalo* meaning great, and *saur* meaning lizard or reptile.

By 1842 so many fossil bones from huge reptiles had been found that Richard Owen, the British paleontologist, created the word *Dinosauria* from the Greek *deinós* meaning terrible. And so the dinosaurs came into being. Owen introduced new dimensions into the understanding of past animal life on earth—and in London some of his work can still be seen in the huge concrete models of prehistoric animals constructed under his direction in the grounds of Crystal Palace in South London. These were made in 1854 when the original Crystal Palace, built in Hyde Park for the great Exhibition of 1851, was reerected in Sydenham Park. On one occasion a dinner party for 20 people was held inside the body of the *Iguanodon* model, which was nearing completion. Owen sat in the head of the dinosaur to make the toast.

And yet although the scientific study of dinosaurs began in Britain, it was America that was to prove one of the world's richest sources of specimens. It was not until 1877, 12 years after the end of the Civil War, that a new kind of war began to rage in the "Wild West." This war was not between cowboys and

Fascinated by the Dinosaurs

Left: Benjamin Mudge, Marsh's chief dinosaur collector, checks Arthur Lakes' discoveries at his dig in Colorado. Lakes had written to Othniel Charles Marsh saying, "Whilst I am thoroughly embued with the enthusiasm attached to such pursuits and discoveries and should greatly like to continue them I have not the pecuniary means to do so."

Indians but between two different groups of dinosaur collectors, each determined to find fossils before the other. The leaders were two scientists, Othniel Charles Marsh, a professor at Yale University, and Edward Drinker Cope, owner and editor of *American Naturalist*. They loathed each other and were arch rivals for two decades, often brawling in newspapers and scientific publications. It is known that two of their paid collectors in the field, on meeting in Wyoming, fought each other with their fists. But without doubt it is due to their efforts that some of the richest dinosaur finds were made.

Arthur Lakes, an unassuming young geologist, sent bones to Marsh with a request for funds to continue his work, but he heard nothing until Marsh learned that Cope had also received bones. Spurred into action, Marsh reported Lakes' work and finds in a scientific journal and sent his chief collector Professor Benjamin Mudge to Lakes' dig in Colorado. He also sent a check for $100 to assist Lakes. The Colorado dig gave out after two years, and a new site was found in Wyoming. The weather proved terrible with gales, sandstorms, ice, and snow. After less than a year Lakes gave up and went to teach geology at the Colorado School of Mines. But he remained bewitched by dinosaurs for the rest of his life and in 1914, when he was 70 years old, he used his diaries and detailed drawings to paint several reconstructed landscapes showing dinosaurs feeding and doing battle. The paintings were imaginative and generally accurate.

Since then many have contributed to the work of unearthing dinosaurs and piecing together what kind of lives these giants led during their 135 million years on earth, until they vanished about 65 million years ago. Today dinosaurs not only fascinate children but they have also been subjected to every kind of science fiction presentation possible, from horror films to novels.

Dinosaurs are fascinating not only because of their immense proportions, but also because they seem to have suddenly disappeared without any real reason. But it should not be thought that dinosaurs were the only large reptiles of their time. Reptiles evolved in the early Carboniferous period around 300 million

Below: Arthur Lakes' painting of an allosaurus in mid-leap, attacking a ceratosaurus. His imaginative work has been preserved in the Arthur Lakes Library at the Colorado School of Mines where he taught geology for many years.

"Guide Fossils"

Right: supposedly a living plesiosaur, discovered by a group of Patagonians in 1922 on an expedition from Buenos Aires. Many people believe the disputed Loch Ness monster is a plesiosaur, one of the group of cold-blooded vertebrates known as amphibians.

Below: a nest of dinosaur eggs found in the Gobi Desert in Mongolia. Specimens of the ceratopsian (three-horned) dinosaur *Protoceratops* at various stages of development were found nearby. The nests, simple holes scooped in the sand, contained rings of thick-shelled eggs. Like modern turtles, the females must have laid their eggs, covered them with sand, and left them to hatch under the heat of the sun.

years ago from the *amphibians*, cold-blooded vertebrates whose young live in the water while the adults are adapted to life on land. Amphibians were a dominant group for the comparatively short span of less than 50 million years, and reptiles proved much more successful than their predecessors in several ways. Their most important survival characteristic was that the eggs of reptiles were enclosed in firm protective shells, each with its own supply of water and food inside. This made reptiles independent of the water allowing them to become a dominant land group, unlike the amphibians who had to return to aquatic environments to lay eggs which were surrounded only by a soft protective jelly. Dinosaurs, being reptiles, laid eggs, and some of these have been recovered from the Gobi Desert in Mongolia among other places. By the Permian period, around 250 million years ago, the heyday of the amphibians was over. By the Triassic period at the opening of the Mesozoic era which began about 220 million years ago, the dinosaurs were evolving, and the oldest known dinosaur fossils relate to the end of the Triassic period. This was the time, for instance, of the 20-foot-long *Plateosaurus*, a large awkward animal which could walk on its back legs and thereby reach high-growing vegetation on which it fed.

The *Lystrosaurus*, an early Triassic *therapsid*—the order which gave rise to mammals—figures in an interesting fossil discovery made in 1969. These heavily-built animals, characterized by a barrellike body and short, strong legs with broad feet, also had a very distinctive head with beaklike jaw and a conspicuous pair of tusks in its upper jaws. *Lystrosaurus* was common in Africa, and its fossils are so numerous in the lower beds of Triassic rock in

South Africa that the beds are known as the "Lystrosaurus zone."

As a matter of fact, they are often termed "guide fossils" by geologists and paleontologists because they are so readily recognizable. In 1969 these fossils were discovered on the Antarctic continent. Here was a land animal of over 200 million years ago which was virtually identical with similar fossil forms found in Africa, India, and China. This evidence that somehow *Lystrosaurus* had reached Antarctica, which was then not the frozen waste we know today but a warm tropical environment, was impressive confirmation of the theory of continental drift (discussed in Chapter 2). This claims that the continents of Africa, India, and Antarctica had been joined at an early stage in the earth's history in the single continent of Gondwanaland. From its structure, it appears that the small hippopotamuslike *Lystrosaurus* was unlikely to have walked hundreds of miles over long land bridges (even if these had existed) or swum across oceans. It was a swamp-dwelling creature, and with the continents connected it probably migrated at a leisurely pace over the swampy areas which had formed during the several million years of the early Triassic period. From the fossil evidence it seems that *Lystrosaurus* traveled in small herds and was a social animal. The creatures also seem to have been constantly accompanied by other early dinosaurs. These were small, socket-toothed reptiles known as *thecodonts*. The companions of *Lystrosaurus* were *Proterosuchus*, who perhaps gained some protection from associ-

Below: a reconstruction by Czech artist Zdenek Burian of *Lystrosaurus*, a dinosaur which bears a certain resemblance to a small hippopotamus. It measured about 4 feet in length and was probably aquatic, although it was also able to walk on dry land. Most of its teeth were modified so as to form a kind of beak like that of a turtle. Onetime resident of Gondwanaland, its fossil presence in Antarctica proves that the continent of the South Pole was once part of a much larger southern land mass.

A Multitude of Dinosaurs

ating with herding animals. This can be seen today in the mixed herds of zebras and antelopes, for example, which graze together in Africa.

During the second half of the Triassic period two distinct orders of dinosaur evolved from thecodont ancestors. These were the *Saurischia* ("lizard-hipped) and the *Ornithischia* ("bird-hipped"), which are distinguished by the differences in the structure of their pelvises or hipbones. Most saurischians had a full set of teeth, and the group included both plant-eating and flesh-eating dinosaurs. The ornithischians, many of which lacked front teeth, evolved beaks instead and were probably plant-eaters only.

The major dinosaur group of the late Triassic period was the *prosauropods*. The majority had bulky bodies with long necks and tails. Most had small, serrated teeth to deal with a diet of plants, although it seems some were flesh-eaters. The herbivorous ones increased in size as they evolved. One type of *Thecodontosaurus* was only 6.5 to 10 feet long and, although it usually moved on all fours, it probably reared up on its hind legs to reach tall foliage or run fast, the long tail acting as a counterbalance. The larger forms, such as *Plateosaurus*, would not have been able to run in this fashion because their 20-foot lengths would have been too unwieldy, although they could probably stand up on their hind legs to feed. The prosauropods had died out by the end of the Triassic period, by which time their successors, the *sauropods*, had taken over. Probably descended from the early prosauropods, these dinosaurs are the ones whose immense size continues to excite our minds; they are particularly impressive when one stands underneath a reconstructed model or skeleton in a museum.

The largest dinosaurs of all, mostly herbivores, lived in the Jurassic period around 150 million years ago. *Brachiosaurus* was over 40 feet high and weighed something like 80 to 100 tons, while *Diplodocus* was over 90 feet long. The lifestyle of the

Right: *Plateosaurus* as restored by G. Biese, showing a conception of the sort of environment in which this dinosaur may have lived and died. Several entire skeletons and related fragments have been found in late Triassic deposits in Germany, France, and South Africa. With its serrated teeth, clawed fingers and toes, and bulky form, the *Plateosaurus* must have been a powerful and terrifying enemy.

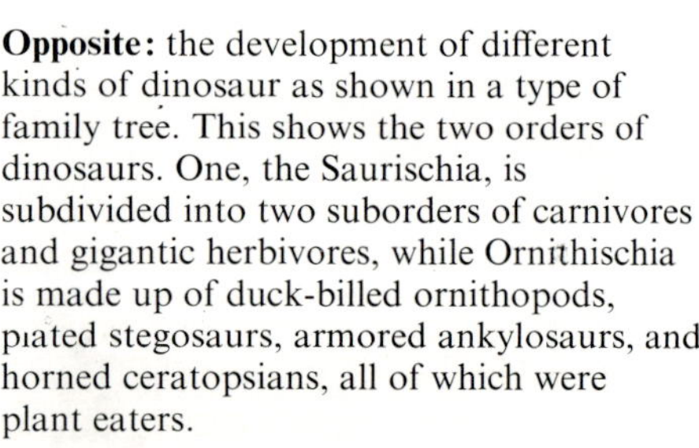

Opposite: the development of different kinds of dinosaur as shown in a type of family tree. This shows the two orders of dinosaurs. One, the Saurischia, is subdivided into two suborders of carnivores and gigantic herbivores, while Ornithischia is made up of duck-billed ornithopods, plated stegosaurs, armored ankylosaurs, and horned ceratopsians, all of which were plant eaters.

Triceratops
Tyrannosaurus
Trachodon
Ankylosaurus
HADROSAURS
CERATOPSIANS
Struthiomimus
Iguanadon
Brachiosaurus
Stegosaurus
ORNITHOPODS
Coelophysis
SAUROPODS
Plateosaurus
PROSAUROPODS
CARNOSAURS
COELUROSAURS
SAURISCHIA
ORNITHISCHIA
Ornithosuchus
PSEUDOSUCHIA
Euparkeria

Were Dinosaurs Warm-blooded?

Right: *Brachiosaurus brancai*, a giant herbivorous saurian (75 feet long and 40 feet high) that lived in Africa, North America, Europe, and eastern Asia during the Jurassic and early Cretaceous periods. As its forelimbs were longer than its hindlimbs (the name means "arm lizard"), it differed significantly in proportion and attitude from both *Brontosaurus* and *Diplodocus*. Its eyes were set high in its head with the nostrils raised on a crown. According to many paleontologists as well as the Czechoslovak artist, Zdenek Burian, who drew this picture, *Brachiosaurus* and the other sauropods dwelled in the water, a medium that would help to support their great weight.

Brachiosaurus is still surrounded by mystery. In many books and scientific papers it is portrayed as a huge amphibious reptile, living in deep lakes or rivers, a description based on the assumption that their short limbs would not have been able to support the weight of their enormous bodies unless they remained submerged so that the water could help hold them up. However, it is difficult to see how these giants could have obtained enough plant material in their watery homes unless they left them to feed on dry land. *Brachiosaurus* would have had to leave the water in any case to reproduce and lay eggs, so its legs must have been able to support the animal on land. However, another school of tnought suggests that a long neck was evolved specifically so that the creature could move through deep water and still breathe air. In fact, the nostril of this dinosaur was located at the top of its head, possibly so that in the presence of enemies the animal could sink, hiding all of itself except the nostril which would still allow it to breathe. But in this case the body of the animal would be subject to such a high level of pressure exerted on it by the deep water that it would have found it difficult to expand its chest and draw air into its lungs. Another interesting point is the shape of the animal's small, unwebbed, elephantlike feet, which would have been useless if the sea or river bottom were muddy—perhaps causing the animal to sink and become trapped. An American paleontologist leads the school of thought which hypothesizes that the sauropods were primarily land animals. Other evidence in support of his case is that the legs of sauropods were like pillars and the rib cage was deep, rather like the modern elephant. Also the tail seems to have shown no adaptation for aquatic life, such as a flattening for use as a rudder or an oar. The majority of paleontologists now believe *Brachiosaurus* and other sauropods were land-dwellers, using their long necks like modern giraffes to stretch up for foliage.

However, one of the newest and most startling theories about dinosaurs is the suggestion that the sauropod giants such as *Brachiosaurus*, *Diplodocus*, and *Brontosaurus* were warm-blooded and had a high energy-producing metabolism. The man responsible is Adrian J. Desmond, a young physiologist who became interested in vertebrate paleontology. He presented his ideas in the book *Hot Blooded Dinosaurs*, published in 1975. Being warm-blooded (*endothermic*) was the only way, he claims, that such huge animals could have managed the sustained energy output they required to carry such great weight and remain on their feet. Desmond also suggests that their mammallike physiology would help to explain their mammallike posture. However, the hot- or cold-bloodedness mystery will probably never be solved.

Living alongside these herbivorous, four-legged giants were many types of carnivorous dinosaurs that fed on young or sick plant-eaters. Some were the size of a chicken, for example *Compsognathus*, which lived in the late Jurassic period. This dinosaur appears to have been an agile little two-footed animal (*biped*), which could run at great speed to catch its prey, probably an assortment of small or plant-eating dinosaurs, small mammals, pterodactyls, and early species of birds, the first remains of which belong to this time.

Left: *Compsognathus*, one of the theropods of the order Saurischia. Its overall length was about 2 feet, and the head was only 3 inches long, set at right angles to a long, slender neck. The tail was longer than the rest of the body put together. The most famous specimen is in the Academy of Sciences in Munich, Germany. The exaggerated curve of its backbone is thought to show symptoms of a serious disturbance of the nervous system that probably caused its death some 150 million years ago.

Below: reconstruction of *Archaeopteryx*, prey of the "bird-robber" *Ornitholestes*. It could not fly but only glide, and as it is a representative of an early stage in the evolution of birds, it still has some reptilian features. The tail was long and bony, and the wings bore claws to help the bird climb. Its scaly head contained a beakful of sharp teeth, but the rest of its body was covered with feathers—evidence that clearly identifies *Archaeopteryx* as a bird. The first fossil skeleton of this creature was found in Bavaria in 1861.

Another interesting dinosaur which probably descended from the same ancestral stock as *Compsognathus* was the "bird-robber" *Ornitholestes*, which was a small, 6-foot-long biped. Its structure has been well known since an almost complete skeleton was found in upper Jurassic layers of a dig in Wyoming. What is fascinating are the forelimbs, which are quite large, with specialized "hands" tapering to long slender "fingers." The first finger is short and opposable so that it may have been used like a human's thumb to allow the dinosaur a good grasp. With these hands many paleontologists think that it caught the birds of its time, such as crow-sized *Archaeopteryx*.

"Flesh Lizards"

With a large population of huge plant-eating dinosaurs, it is not surprising that giant meat-eating dinosaurs evolved to take advantage of the plentiful food supply. These were the *carnosaurs*, or "flesh-lizards," which evolved at the beginning of the Jurassic period. The 30-foot-long *Megalosaurus* was typical of this group with its large head, bipedal movement, short neck, long body, and long, powerful, balancing tail. The hind feet were birdlike.

However, the most famous carnosaur was to appear in the Cretaceous period, which began 135 million years ago. The *Tyrannosaurus* was extremely large for a killer—at least 45 feet from its snout to its tail and towering 16 feet above the ground. The 4-foot-long skull was deep and big jaws contained 3- to 6-inch-long, daggerlike teeth with serrated edges. This is the monster which has been portrayed in endless horrific monster movies. Although a ferocious, terrifying animal, it is still somewhat a mystery why its forelimbs shrank to almost useless appendages (although they were still almost 3 feet long). Most paleontologists believe that it battled with its jaws in a slow, clumsy fashion, finding and killing its prey by instinctive behavior.

The second major order of dinosaurs, the ornithischians, appeared after the saurischians in the late Triassic period. First were quite small 3-foot-long species such as *Fabrosaurus*, which even at this stage had the horny beak typical of this order. The early forms, such as *Camptosaurus* of the Jurassic and *Iguanodon* of the early Cretaceous period, were still bipeds. *Camptosaurus* varied from 6 to 20 feet long, with strong hind limbs on which it could be quite active. It appears from its skeleton that the front legs were strong enough to allow the creature to come down on all fours when it wished. *Iguanodon* was over 30 feet long, and a great deal is known about it. This is because, apart from the usual bones, teeth, and fossil skeletons of various ages, skin impressions and footprints have also been discovered. The pioneer work

Below left: a Jurassic delta landscape, habitat of the flesh-eating dinosaurs. Dominated by gymnospermous trees, such as the swamp cypress and maidenhair tree, there were as yet no flowering plants. It was a period of luxuriant vegetation after the aridity and barrenness of the Triassic period.

Below: *Iguanodon*, Dr. Mantell's bipedal, herbivorous dinosaur. Its skull is compressed from side to side, and the large nostrils are at the front, near the tip of the skull. The hands have five fingers, but the thumb consists of a spikelike structure that was originally thought of as a rhinoceros horn, and which in some early reconstructions of *Iguanodon* was placed on the nose. It took many years, in fact, to get it off the nose and onto the hand.

done by Dr. Gideon Mantell was built upon and a great advance made in the knowledge of this huge beast with a celebrated find in Belgium in 1878. Coal miners were enlarging a mine shaft some 900 feet below the surface at Bernissart when they found a collection of large bones which the Natural History Museum in Brussels later identified as *Iguanodon* remains. Not only was there one skeleton, but several. It seemed that a large number of the animals had met their death by falling into a ravine, later becoming trapped in the deep mud at the bottom. Eleven complete skeletons were found as well as another 20 which were incomplete. More than one species of *Iguanodon* is now recognized. *Iguanodon bernissartensis* stood with its head about 15 feet from the ground and the length along its backbone was 30 feet. The jaws had large leaf-shaped teeth at the side, but there were no teeth at the front. However, special bones covered with horn were excellent for snipping.

Food gathering must have been greatly facilitated by its five-fingered hands. Mantell had suggested a mysterious horn on the nose of his *Iguanodon*, but the Belgium finds showed the horn to be actually a thumb bone. At rest this dinosaur squatted on its tail as do modern kangaroos. When running, the tail was raised off the ground and the head bent forward so that the animal was

Right: *Iguanodon* skeleton found deep below ground in a Belgian coal mine at Bernissart near the French border in 1878. Because of the difficulties under which the miners worked—in a small tunnel at a great depth with only very dim light to guide them—it appears that they picked their way right through one skeleton and had almost completely destroyed it before they became aware of its presence. P. J. van Beneden, a noted Belgian paleontologist of the day, was sent to examine the bones and declared them to be *Iguanodon* remains.

Plated Reptiles

balanced. Considering it probably weighed over 7 tons, this animal could reach quite a speed.

The "bird-hip" dinosaurs were characterized by amazing bizarre decoration and armor, but in general they remain a mystery today. Among the different forms were duck-billed dinosaurs with broad, flattened snouts that shoveled up food from the muddy bottoms of streams and lakes. And although we can only guess that certain dinosaurs were aquatic, it seems definite in this case because of the webbing between the toes. However, the fossil remains of their stomachs suggest they ate land plants, so perhaps their duckbills were used for stripping.

Certain of these dinosaurs stopped being bipedal and began to walk on all fours, thus becoming much slower, but it seems that the development of armor compensated for this loss. One of the earliest ornithischians to have been found—*Scelidosaurus*—was armored. Remains unearthed in Dorset in southwestern England show that it was some 12 feet long and stood low to the ground. It was covered by longitudinal rows of bony plates or *scutes* as well as small spikes; those on the muscular tail were no doubt effectively used as weapons.

The best known of this group was *Stegosaurus* ("plated reptile"), which grew to a length of 20 to 40 feet. The head was low to the ground while the body was highly arched with paired, alternately arranged, triangular bony plates down its length. These were largest at the highest point of the back, where they were almost 3 feet across. The last few feet of the tail had a double pair of spikes, and these could be useful weapons when swung sideways against an enemy. Although certain paleontologists have interpreted the plates as defense against the carnivorous dinosaurs of the period, it now seems that this may be a wrong hypothesis. The plates were implanted in the skin with no connection or support from the axial skeleton. A large carnivore

Above left: *Stegosaurus*, "plated reptile," which has a long, narrow skull and a horny beak. The brain cavity was relatively small and indicates a small brain of low organization, one of the smallest in fact of the known land vertebrates. Professor Othniel Charles Marsh made the first picture of its restored skeleton in 1891. As its weight must have been around 2 tons, agility was not one of its strong points.

Above: the Canadian duck-billed hadrosaur *Hypacrosaurus*, which was 25 feet long and walked, like most bipedal dinosaurs, with its tail outstretched. Hadrosaurs, large foliage browsers of the late Cretaceous period, were an extremely successful group and flourished in great numbers in North America. They are therefore some of the best-known dinosaurs.

A Second Brain?

could probably attack quite easily, especially at the front where the spiny tail could not reach. Perhaps the plates were simply a warning signal to other animals, just as a cobra raises its hood in a display of threatening behavior. A recent suggestion is that, since they were permeated by blood vessels, the plates could dissipate excess heat like modern radiators and thus provide an effective cooling system.

One of the most interesting facts about *Stegosaurus*, however, is that it had two brains! This dinosaur possessed a small, walnut-sized brain in its head as well as a marked enlargement of the stout spinal cord nerves from the legs, which has been interpreted as a second brain. A myth arose that *Stegosaurus* was blessed with two brains, as suggested in the poem "The Dinosaur" by the late Bert Taylor, which says in part:

> You will observe by these remains
> The creature had two sets of brains—
> One in his head (the usual place),
> The other at his spinal base.

However, it is usually thought that the hind "brain" served merely to help the back legs perform their necessary powerful, pushing functions, although recently another theory has been put forward. Since other types of dinosaur also possessed a similar enlargement of the spinal cord, it has been suggested that perhaps this did not house nerve ganglia at all. The cavity may have contained a gland that secreted *glycogen*. This sugarlike substance, also called animal starch, is readily converted into energy and may have acted like adrenalin to quicken responses when the dinosaur needed to run away or defend itself in the face of danger. The modern ostrich has a similar gland structure in the same part of its body for just such a purpose. The stegosaurs were widely distributed, their remains having been found in North America, Eurasia, and Africa, but they must have failed to adapt somewhere along the line because by early in the

Below: dinosaurs of the late Cretaceous period in a painting by Charles Knight. The peak of dinosaur evolution was reached shortly before the extinction of the group. Prominent in North America 90 million years ago were duck-billed dinosaurs such as *Edmontosaurus* (right) and the hooded *Corythosaurus* and crested *Parasaurolophus* in the swamp on the left. Other contemporary dinosaurs were the heavily armored *Paleoscincus* in the center and the ostrichlike *Struthiomimus* in the background of the painting.

Above: skin imprint of an unarmored dinosaur, the duck-billed *Anatosaurus*. Unlike the armored dinosaurs such as *Scolosaurus*, the skin of the unarmored creatures is imperfectly known. Here it is shown to be thin and flexible, with a mosaic pattern of small polygonal plates. This fossil was found in North America.

Cretaceous period they had become extinct (perhaps they would have proved more successful if they *had* evolved two brains).

At the beginning of the Cretaceous period, however, the true armored dinosaurs or *ankylosaurs* ("curved reptile") appeared. Even the fossils of early specimens of this group show them to have carried armor—*Acanthopholis*, for instance, had bony plates along its back and sides. By the late Cretaceous period there were a variety of types of ankylosaur, some quite large. Fossils of *Ankylosaurus*, whose name is used to denote the entire group, have been found in Montana and in Alberta, Canada. This massive creature was about 17 feet long, and its overarmor breadth was almost 6 feet. Its weight, which must have been considerable, was supported on short, massive legs which probably ended in huge, elephantlike feet. The weight of the armor would have put terrific stress on the backbone and ribs, so the widely-spread ribs and the vertebrae were particularly hard for additional support. Its plates would have protected *Ankylosaurus* from its enemy *Tyrannosaurus*; the only unprotected areas were its sides and belly. It is hard to imagine any *Tyrannosaurus* rolling over on its back in order to attack the ankylosaur in its vulnerable parts.

Fossils of armored dinosaurs are quite numerous on the North American continent. The skin of one armored dinosaur, *Scolosaurus*, is particularly well documented because its imprint was discovered in sandstone near the Red Deer River in Alberta. Its back was covered by a series of paired polygonal plates and spines of different sizes. Hundreds of little *ossicles*, or small

The Horn-faced Dinosaurs

Below: excavating a fossil plesiosaur, a large aquatic reptile that flourished during the Jurassic period. Bones grouped as in hands and feet represent the ends of limbs that had developed into flippers to propel this long-necked, sharp-toothed fish eater. When the surface of a bone, skull, or vertebra to be excavated has been more or less exposed, great care must be taken in moving it. It may already be fractured, so pieces of fine paper are dampened and placed over the bone; upon these are placed strips or bandages of cheesecloth or burlap soaked in liquid plaster of Paris. When these set, the bone may be turned over in order to work on the other side.

bones, were embedded in the rest of the skin which must have made it more flexible. Its name means "thorn reptile" and comes from the animal's resemblance to the skin of the thorny devil or moloch, a modern lizard of the Australian desert region. Because the first *Scolosaurus* fossil was found in a layer of sandy rock, it was suggested at the time that the animal must have devoured insects as little else was believed available. An animal 18 feet long would have had to eat a minimum of 7000 grasshoppers and beetles a week in order to survive, and it would have been the largest insectivore on record. However, other fossils were later found in a region of swamps and deltas where vegetation abounded, which seemed to indicate that the heavy, slow-moving *Scolosaurus* fed on the vegetation and later wandered into the nearby sand dunes. That first fossilized specimen was probably injured when a sandbank gave way under it. The *Scolosaurus* would have rolled over on its back, completely helpless and unable to move its 2-ton bulk, and probably died a slow death by starvation. No doubt if a *Tyrannosaurus* had come across the creature with its soft belly exposed the predator would have relished the easy meal.

The final group of dinosaurs to emerge in the Cretaceous period were the ceratopsians or horn-faced dinosaurs. They have been found only in North America aside from a few specimens such as eggs and embryos in Mongolia. The simplest early types

were less than 6 feet long and had small, bony frills protruding from their necks. They may have evolved from bipedal ancestors. In later specimens the frill was enormous and projected far over the backbone, and these creatures had also increased in size to 30 feet in length—of which the skull alone accounted for over 8 feet. Although some horned dinosaurs did not actually have horns on their skulls, others such as *Triceratops* had up to three horns.

Protoceratops, about 5 to 6 feet long, was the most primitive of all the horned dinosaurs so far found, although they are known to have existed only from the later Cretaceous period. In the mid-1920s some eggs were discovered in the Gobi Desert in Mongolia. The eggs were found along with the remains of very young dinosaurs as well as adults. When the jigsaw of bones was put together it seemed apparent that the other dinosaurs might have meant to steal the eggs or suck out the contents. The 6-inch-long eggs, which were ellipsoidal in shape, were arranged in a spiral and almost circular manner. This was probably so they could develop in the sand in the sun's heat. The adults found so close to the eggs raises an interesting question: did the *Protoceratops* show parental interest in their offspring? The attitude of modern reptiles such as the crocodile and alligator is unclear, in that the females guard their nest and on hearing their squeaks even help the hatching young to scramble out of it by uncovering the protective layers of vegetation. However, they do not show much parental care after this stage, and in fact many offspring are gobbled up as food by the now unconcerned parent. These huge, rhinoceroslike animals did not rely on armor or nonaggressive tactics for defense as did the tanklike ankylosaurs. They probably charged attackers with their horns instead, as do the herbivorous rhinos of our times who are basically placid unless provoked by aggressors. *Triceratops*, at 8.5 tons and some 36 feet in length the largest of the ceratopsians, must have been an awesome sight in battle with *Tyrannosaurus*. When attacked this creature would face its enemy with head lowered so that its horns were ready to impale the other's body.

However, no matter how large or small the dinosaur, no matter how aggressive or well equipped for defense against other dinosaurs, by the end of the Cretaceous period 70 million years ago not one was left in any part of the globe. Why and how were these mighty beasts defeated? The question crops up again and again, and every few years a new hypothesis is put forward. Some of the older suggestions are rather amusing: the dinosaurs were overtaken by suicidal psychosis, and all marched to their doom; little green men in flying machines arrived on earth and killed them all; Noah did not have enough space in the Ark for two of these massive creatures.

However, an important point to remember is that it was not just the dinosaur that had disappeared by the end of the Cretaceous period. Also on the way out were the batlike, flying pterosaurs of the air as well as the paddle-footed plesiosaurs and sea-serpentlike mosasaurs of the oceans. Certain invertebrate groups also disappeared suddenly, such as the ammonites (a type of mollusc) and certain foraminifera (microscopic shelled marine animals). Yet certain other groups survived—croco-

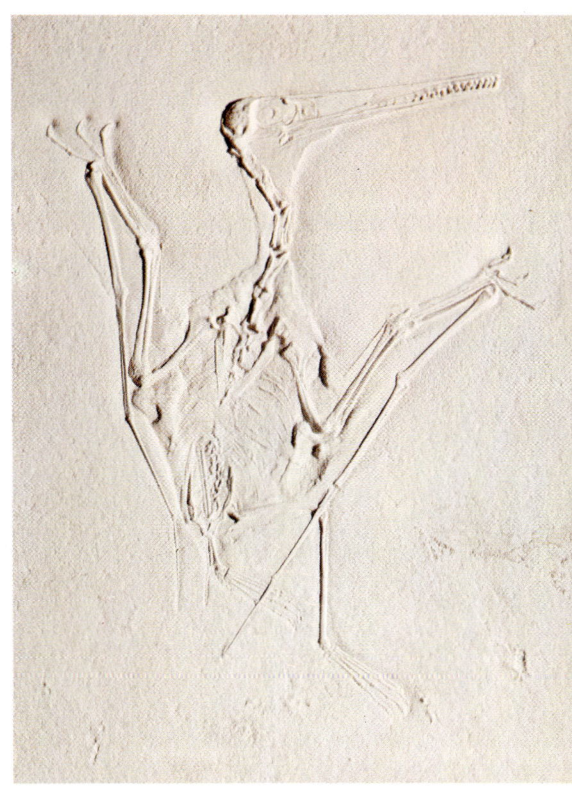

Above: pterodactyl fossil (a pterosaur) from the upper Jurassic limestone of Bavaria. In 1784 the first such fossil was found at Eichstätt by the Italian naturalist Cosmo Collini, and it was these fossil remains that helped prove that whole races of animals (dinosaurs among them) had disappeared from the face of the earth. Extinction of archaic primal life was thus established as a fact of life.

What Happened to the Dinosaurs?

dilians, turtles and their relatives, snakes, mammals, and birds as well as numerous types of invertebrates.

Hence it is necessary to look at the earth and seas as a whole and to survey the various factors that might have affected all the animals that became extinct. One reason often given is that the plant life changed profoundly. This meant that the vegetation available for food for the herbivorous dinosaurs was not appropriate for them, so that they died out through starvation. Naturally the flesh-eaters would then have died through lack of prey. However, flowering plants (*angiosperms*) had evolved and become dominant over the previously abundant pines and other *gymnosperms* some 35 million years before the extinction of the dinosaurs. In fact the wide variation in plant life may have been a contributing factor allowing the dinosaurs, particularly the hadrosaurs and ceratopsians, to evolve into so many different forms during the Cretaceous period.

Another theory suggests that the newly-evolved flowering plants were the agents of a sort of "chemical warfare" that wiped out the dinosaurs. Its adherents claim that the earliest angiosperms produced poisons—potent alkaloids, most of them very bitter and some, such as strychnine, highly toxic. These chemicals were evolved to be distasteful and thereby repel any feeding animal, but ingested in huge quantities the concentration would be lethal. The herbivorous dinosaurs would probably have been unable to detect the low concentration of poison, but with their tremendous appetites the poison would eventually build up within their bodies to the point at which it would prove fatal. This process has occurred in modern times, when man has sprayed strong insecticides and pesticides on crops to increase their yields. These substances have moved through the food chain to build up in certain animals, such as birds of prey, to the point where they have wiped out great numbers and reduced some species to near extinction. However, this theory in regard to the dinosaurs seems an unacceptable solution in that flowering plants had evolved in the early Cretaceous period (about 120 million years ago) and had coexisted for 50 million years with the dinosaurs, who did not reach their peak until 30 million years later.

It is now generally accepted that the end of the dinosaur world was not due to any of the foregoing reasons. It is also improbable that the small mammals, which had coexisted with them for 135 million years, could have in some way been the major cause of their decline as they had shown no previous signs of incompatibility.

The most plausible theory is that climate was a major factor. There is no evidence that any Ice Ages occurred during the Cretaceous period, but it does seem that very cold conditions developed which affected the whole world at this time. These conditions may have been seasonal, and those animals that were unable to adapt to the changing climate would have been particularly vulnerable. Both land and sea were affected. The surface waters of the oceans would have cooled quickly if the sun's heat were blocked by dense cloud, while the deeper, colder waters would have been only slightly affected. Indeed, it is true that the animals which disappeared from the seas were those who lived in the surface waters. On land, the cooler seasons affected those animals who were unable to keep their own bodies warm. The warm-blooded animals of the world at this time, the small mammals and birds, had an insulating layer of hair or feathers to keep the temperature high within their bodies. Dinosaurs, however, had naked skins which allowed the heat inside their bodies to escape, so that whatever their size they were doomed. When the cold came, the cold-blooded snakes, tortoises, and other land animals could probably burrow in the ground or find a tree hole, crack in a rock, or a nest to protect them. This was impossible for the enormous dinosaurs.

What caused the cold is still a mystery. One dramatic suggestion is that a star exploding into a supernova caused the weather changes. It could have happened at such a great distance from the earth that the cosmic radiation produced would have been

Below: the shrewlike *Phascolotherium* of Jurassic times, compared in size with dinosaur eggs. An early mammal, this creature had already evolved by the time the last of the great dinosaurs became extinct. The small mammals then took advantage of their new opportunities for food in the Cenozoic era, but they found that wide seas blocked their way to many parts of the world. Thus the separate continents came to contain different groups of mammals.

The Mystery that Remains

absorbed by the earth's ozone layer or ionosphere, so that little lethal radiation would have affected life on the earth. However, great disturbances would have occurred in the climate. Water vapor in the upper atmosphere would have condensed into ice crystals and acted as a screen to prevent the radiant heat of the sun from reaching the earth, thus causing a reduction in temperature. One possible flaw in this theory is that the effects on earth of a supernova would be quite sudden and should have shown up in the fossil record, whether over a period of days, weeks, months, years, or even centuries. But the extermination of the dinosaurs and other groups was a gradual process reaching a sudden conclusion at the end of the Cretaceous period, so there are still many gaps in this theory which need to be filled in.

Although the why is still a mystery, we do know that the dinosaurs were under considerable stress at the end of the Cretaceous period. Dinosaur eggs from the beginning of the late Cretaceous period had thick shells—up to a tenth of an inch. It has been discovered that dinosaur eggshells from later times were much thinner, often only 4 hundredths of an inch in thickness—so that they must have been extremely fragile. This alteration is frequently the result of stress, as modern birds are also known to lay eggs with much thinner shells under conditions of poor food, chemical poisoning, overcrowding, or cold weather. Thin-shelled eggs are formed because the delicately-balanced hormonal system which is responsible for the eggshell's thickness as well as the number of eggs laid is disturbed. Thus stress would not only have affected the adult dinosaurs, but would also be disastrous for the offspring, some eggs cracking and others lacking sufficient cal-

Right: *Leptolepsis*, a 9-inch-long Jurassic fish. It had jaws, a backbone consisting completely of bone rather than cartilage, and other characteristics that help to place it among the earliest of the *teleosts*, or bony fishes. Late in the Cretaceous period the teleosts began to become predominant among fishes, and afterward they evolved along many different lines and adopted a wide range of habits through adaptive radiation. They now include most kinds of living fishes.

cium in the shell for the dinosaur embryo to absorb for the developing bones of its body. Many of the young probably died before hatching, and a few of these were fortunately mummified as fossils, allowing man to study them and hopefully thereby solve the mysterious riddle of the fate of an entire population of enormous creatures who were decimated in the space of only 5 million years.

Above: fossil crocodile, ancestor of the only two living members of the once dominant group of "ruling reptiles" or archosaurs—modern crocodiles and alligators. Their grotesque appearance and somewhat aggressive habits have changed little since they first evolved during the late Triassic period. Lurking by the banks of lakes and rivers, they prey on any animals which may come to the water's edge to drink. They escape detection by lying motionless and almost completely submerged, with just their eyes and nostrils rising above the surface. The side-to-side lashing of their powerful tails makes them strong swimmers, but on land they are usually only able to move quickly when frightened.

Chapter 5
Prehistoric Survivors?

Living fossils—not a contradiction in terms, but real creatures—are the last remnants of animal families of which most members have become extinct. How have such creatures as king (horseshoe) crabs and lamp shells managed to survive for some 500 million years without evolving any modern adaptations? An even more interesting problem for zoologists is how these animals fit into the evolutionary pattern, and whether any can be classed as ancestral forms linking two separate modern animal groups. One of these relics may yet be found to throw light on the complex mystery of the evolution of life on earth millions of years before the evolution of primitive man.

Oval-shaped, tanklike king crabs thunder over sandy tropical beaches; young chicks with claws on their wings clamber through the branches of their South American jungle home; and heavy, lobe-finned fishes swim in the oceans off southeast Africa. This is not a reconstruction of a prehistoric scene—these animals are living today, examples of what are called "living fossils." They all evolved millions of years ago but, while the rest of their close relatives have vanished as evolution has taken its natural course, these creatures have survived into modern times.

King crabs, sometimes called horseshoe crabs, are not actually crabs but are more closely related to spiders. The rounded dome, ranging from brown to dark olive green in color, is hinged to a hard and roughly triangular abdomen which ends in a long, movable tail spike. It may be up to 2 feet long, and when it is turned over, the dome or *carapace* can be seen to protect a series of pairs of jointed legs.

The fossil record of its class, the *Merstomata*, extends back to the Ordovician period some 500 million years ago. All members of the group, which included giant aquatic 9-foot-long arthropods, are now extinct except three genera and five species. The best known of these is the king crab (*Limulus polyphemus*), found near the northwest Atlantic coast, in the Gulf of Mexico, and along the east coasts of Asia from Japan and Korea south through the East Indies and Philippines. This distribution in two widely separate areas of the globe suggests that they may be relics of a more widespread group, rather than members of one

Opposite: the king or horseshoe crab (*Limulus polyphemus*), member of a small but important group of marine animals once classified with arachnids but today considered a separate class. The king crab has survived almost unaltered since the Silurian period. The shelled body terminates in a long, stout spine which the animal uses to push itself into the mud and to right itself if it becomes inverted. Adult king crabs live on the sea bed where they feed on a wide variety of animal and plant material. Their larva, about half an inch long, bear a superficial resemblance to the extinct class of trilobites.

The Possible "Missing Link"

Right: fossil king or horseshoe crab. These are not crustaceans as are other crabs, but more like arachnids (related to scorpions and spiders). Adult horseshoe crabs today live on the bottom of the sea where they feed on seaweed, mollusks, and worms. Their mouths are on their undersides, surrounded by the legs, and the base joints of the legs have spiny protuberances used to chew their food.

which is still evolving and spreading its range, and fossil evidence from the Upper Jurassic strata in Bavaria, southern Germany, supports this claim.

Why should these creatures survive at all, one may ask? No one can say with certainty why they have been able to do so, with no modern improvements or specializations, in competition with more highly developed aquatic arthropods such as the numerous modern species of crab, lobster, and their relatives. Their survival probably results from a combination of unobtrusive habits and the heavy, hoodlike armor protecting the body and legs. It is believed that the group to which the king crabs belong may have been the ancestors of the *Arachnida*, the very numerous class comprising the spiders, scorpions, mites, and ticks. The king crab may also be a living link with the trilobites, an extinct group of tiny marine animals, because many scientists think the trilobites were ancestors of not only the king crabs but the entire giant class of arthropods. Evidence for this has been found in the development of king crab embryos as well as in the fossil structure of many animals considered transitional or intermediate between the two. Thus living fossils are exciting in that they reveal many facts about bygone forms of life.

An ancestral form linking two modern groups of animals, showing clear intermediate details in structure, has not yet been discovered. However, one animal that comes closer than any other to being the "missing link" between two phyla is the class of onychophorans, of which a common genus is the *Peripatus*.

This is a velvety caterpillarlike creature found in moist places such as under logs in tropical forests, mainly in Australia, Africa, Asia, and South and Central America. It is quite rare and occurs only in certain local regions of these widely separated parts of the world. This fact suggests that it may have been more successful and widespread in the past and that, like the king crab, it is gradually disappearing. The animal has many structures similar to those of the phylum *Arthropoda*, which includes insects, crustaceans, spiders, and centipedes, but it also has some similarities with the phylum *Annelida*, which includes the earthworms, bristleworms, and leeches. The peripatus also has some special features of its own.

It is not surprising that animals exist with structures peculiar to two different phyla—this should occur naturally due to the continuous nature of the process of organic evolution. It is more difficult to determine, however, in exactly which order to place peripatus. Some authorities class it with the arthropods, while the majority consider the onychophorans a phylum of their own. Their name means "claw-bearing," and refers to the curved claws of their feet. Some 65 species of onychophorans are still extant, but all are ancient and do not appear to have changed greatly since the Cambrian period, some 550 million years ago. The only fossil peripatus so far found was discovered in Cambrian rocks.

The shape of the peripatus is that of a long slug with legs, and in fact it was thought to be a mollusc when it was first discovered in 1825. The more or less cylindrical body ranges from a half inch to 6 inches in length depending on the species, and the number of legs varies from 14 to 43 pairs. Unlike typical annelids and arthropods, the peripatus shows no external segmentation, although it has a pair of legs for each internal segment of its body. The legs end in claws similar to those of arthropods, but unlike arthropods the legs are not jointed.

A link with the annelids is the similarly thin *cuticle*, the layer of cells which protects the epidermis below. Its velvety texture in the peripatus is, however, a unique feature. The function of this

Below left: *Peripatus*, a caterpillarlike velvet worm. It lives in moist habitats, under stones or the bark of fallen trees, as it is highly susceptible to drying out. In an ordinarily dry room *Peripatus* can lose a third of its body weight in four hours through dehydration. Its defense mechanism consists of two slime glands at the sides of its mouth which secrete a milky fluid when the animal is disturbed. This substance shoots out for a distance of 3 to 12 inches, congealing on contact with air to form sticky threads which entangle its enemy. Velvet worms are nocturnal and carnivorous, feeding on small insects such as termites and on such crustaceans as woodlice. They also eat the carcasses of large insects like grasshoppers.

Below: middle Cambrian fossil example of *Aysheia*, a caterpillarlike animal which resembles the modern *Peripatus*. However, *Aysheia* was a marine creature. This fossil was found in the fine dark shale of Burgess Pass, high up in the mountains of British Columbia, in western Canada.

Above: a lamp shell, one of the most common of all fossils. Like some mollusks, brachiopods feed by drawing water into their two-valved shells and straining out the food particles.

Below: live lamp shells, showing their stalks. In the past they were probably as numerous as mollusks are today, and in some parts of the world whole layers of rock are made up of little else than fossil lamp shells.

cuticle is probably to prevent the epidermis from becoming soaked with water. Beneath the epidermis are layers of muscles which are also similar to annelid structure. The head of the peripatus is made up of three segments, thought to indicate a midway stage between the annelids and arthropods, the latter having a five-segment head. The internal anatomy is a mixture of annelid-like and arthropodlike structures: the excretory system is most like the annelids' in character, while the respiratory system is most similar to the arthropods'.

It is still something of a mystery why the annelids and arthropods have gone on to evolve into large and successful groups while the onychophorans are probably well on their way to extinction. As with so many questions in the world of nature, a conclusive answer cannot be given. A combination of several factors usually determines whether a group is a success or failure. The factors include characteristics which are inherited as well as others determined by changes in climate or environment; still others are accidental. The peripatus appears to be well adapted to its environment. Water is not easily lost through the dry skin, its body metabolism enables a high production of water, and it is equipped with mechanisms which decrease the amount of water lost through excretion. The animal is *viviparous*, meaning that it gives birth to live young. However, its respiratory tubes, which end in tiny blowholes known as *spiracles* all over its body, are thus always exposed and cannot be closed. As a result the peripatus does not have much control over water loss through these openings. This may be the main reason why an apparently well adapted animal has failed to progress along an evolutionary course. However, the peripatus is still without doubt our best example of a connecting link in the animal kingdom, and should perhaps be given credit for surviving as long as it has.

Another living fossil of the animal kingdom is the lamp shell of the genus *Lingula*. Belonging to a small but geologically important phylum, the *brachiopods*, it has left us an abundant and unusually complete fossil record. Modern species of *Lingula* are almost identical to those which lived around 500 million years ago. In fact, no other animal genus has such a long unbroken history. Only about 300 species of brachiopod are alive today, but at least 30 thousand different species have been identified as once living in the seas of the Paleozoic and Mesozoic eras. In some parts of the world whole layers of rock are composed solely of fossil lamp shells, providing valuable information as geological indicators.

Looking rather like small clams, these marine animals are rarely more than 2 inches long. They acquired their name because some species of lamp shell look like an ancient Roman oil lamp. A pair of shells protects the soft parts within, and a short, fleshy stalk usually sticks out at the rear of the shell to anchor the animal to a rock or coral. Certain lamp shells have a muscular stalk which can be shortened or lengthened to use for burrowing. These dozen or so species burrow in the black, smelly mud of the Indo-Pacific region, particularly off the coasts of Japan, southern Australia, and New Zealand.

Lamp shells were some of the earliest brachiopods and have persisted unchanged through the evolution of the animal kingdom while other groups have come and gone. One reason for their survival is that they can live in foul water which few other animals can tolerate, so they have had few competitors for the available food and very few enemies.

From the invertebrates, which include the living fossils so far discussed, follow on the major group of *vertebrates* or *chordates*, animals with backbones, which developed about the beginning of the Cambrian period some 600 million years ago. The ancestry of these early chordates, which had no true backbone but instead a *notochord*, a thick stiffening rod which ran the length of their bodies, is still a mystery. Highly specialized "leftovers" include sea squirts, acorn worms, and lancelets. They are probably related to the *echinoderms*, which include the sea urchins and starfishes, whose ancestors were the forerunners of the first class of vertebrates. These were the primitive jawless fishes which appeared on the scene about 440 million years ago at the end of the Ordovician period. Today survivors exist in the eellike hagfishes and lampreys of the class *Agnatha*. The hagfishes are far from being the most attractive of fishes, either in looks or habits. They bore into the flesh of dead and dying fish with their horny teeth and use their sucking mouths and rasping tongues to feed on the interior parts. The 21 species that we recognize today have retained the basic body organization of the early vertebrates.

The cartilaginous fishlike vertebrates of the class *Chondrichthyes* were the next group to evolve and include skates, rays, and sharks as well as the chimaeras, a group of particularly grotesque fishes. These bony, fishlike vertebrates had appeared by the end of the Silurian and beginning of the Devonian period, about 400 million years ago. The surviving species of this group are of fairly recent origin and make up only a small percentage of living marine animals. It is in the other main group of fishes that we find our "living fossils"—the class *Osteichthyes*, the bony fishes, which includes the great majority of food and game fishes.

Specialized "Leftovers"

Below: eel-shaped lampreys, creatures without bones but with a skeleton of cartilage. Although they have no jaws, their mouths hold batteries of horny teeth with which they rasp the flesh of living or dead fishes like the carp shown here. There is little competition for this way of life, which may be why the lamprey has survived unchanged since Carboniferous times.

Above: fossil shark *Cladoselache*, a primitive chondrichthyan fish (which class includes sharks, rays, and their relatives) found in the Cleveland Shales of Ohio. It was about 3 feet long with paired fins that were simple triangular flaps. Rows of sharp teeth show it was a predator. One important difference between *Cladoselache* and modern sharks is that the latter have small eyes and hunt almost entirely by smell, whereas *Cladoselache* had large eyes and a short snout. It must have hunted mainly by sight.

Very few of the subclass Crossopterygii have survived to modern times, but they are at the bottom of the evolutionary line that gave rise to air-breathing vertebrates. However, one survivor, thought to have become extinct over 70 million years ago, is renowned for having been found alive in the Indian Ocean. This is the coelacanth, *Latimeria chalumna*, which was pulled up in a local fisherman's net from a depth of 220 feet off the east coast of southern Africa on December 22, 1938. The skipper, Captain Goosen, realized that the 5-foot-long blue fish with strange scales and fleshy-lobed fins was unusual, and he sent it to the curator of a South African museum. She wrote to Professor J. L. B. Smith, a chemist by profession and ichthyologist in his spare time, who because of Christmas festivities could not travel the 400 miles from his home for several days. By that time decay had truly set in, but nonetheless he realized that the fish was a priceless scientific treasure. He brought it to the notice of the scientific world, and the popular press took it up as well.

For the next 14 years Professor Smith carried out a well-organized campaign, advertising and searching for another specimen. World War II interrupted serious searching, and it was not until December 29, 1952 that a second fish was caught, this time off the Comoro Islands near Madagascar. It was caught on

The Amazing Coelacanth

Above: Professor J. L. B. Smith posing with the second coelacanth on Pamanzi Island in the Comoros in 1952. The word coelacanth —pronounced "seelakanth"—means hollow spine. Smith considers the importance of this discovery to lie in an implicit warning to scientists "not to be too dogmatic," as well as a confirmation of the accurate deductions of paleontologists about the structure of coelacanths from fossils.

a hook and line by Ahmed Hussein, a fisherman, who no doubt struggled quite hard to land the 100-pound fish from a depth of about 65 feet. Smith flew in on a plane placed at his disposal by the then prime minister of South Africa, Dr. Malan. Again the specimen had been attacked by bacteria, causing much decay, but "Old Four-legs," as Smith nicknamed the coelacanth, was definitely no longer a one-off phenomenon.

The French then took up the search and, under the direction of Professors J. Millot and J. Anthony, some 70 specimens of these very primitive survivors of the Cretaceous period have since been caught and investigated in detail. But why is the coelacanth so mysteriously interesting?

First, it is the surprising "four legs" that attract the eye. These are actually paired pectoral fins, but the front ones, instead of projecting straight out from the body—as for example on a goldfish or herring—are carried on scaly, muscular lobes and seem halfway between a normal pectoral fin and the walking limb of a primitive prehistoric land animal (such as the early amphibians). The dorsal and anal fins are also lobed in a similar fashion. The tail fin is rather strange, not developing out of a thin constriction as on a goldfish but resulting from a rapid and even narrowing of the body, which continues as a narrow strip dividing the rays of

The Riddle of the Lungfish

the tail fin into two equal parts. The scales, too, are archaic, each a bony plate covered with dermal *denticles*—small, toothlike points on the skin—like those of sharks.

Many internal peculiarities also point to the ancient ancestry of the coelacanth. The heart is very simply structured, even compared with the hearts of other living fishes, and it is similar to what is considered by comparative anatomists to be an early stage in the evolution of the heart. Many scientists had expected the coelacanth to have an air bladder, whose function would not be to increase buoyancy as in bony fishes but to be a primitive lung. If this had been so, the creature would have been seen as a very early "four-legged," air-breathing link between the sea-dwelling animals and those that crawled onto the shore to become the first amphibians. Alas the scientists were disappointed, since although coelacanths did possess an air bladder, it was slender, rigid, and filled with fat, so that its function could be neither respiratory nor even hydrostatic. As yet its function remains a mystery.

Using modern preserving techniques of deep freezing freshly caught specimens, scientists have been able to study the physiology of this ancient survivor in greater detail. They have found a high content of urea in its liver, suggesting that *Latimeria*'s process of nitrogen metabolism is similar to that of sharks, lungfishes, and amphibians but unlike that of the modern bony fishes. Eggs have also been found which are quite large—3.5 inches in diameter—and few in number. Only 19 were found in one 5-foot specimen. They appear to be unprotected, nor is there any sign that they complete their development inside the mother. Perhaps the eggs are protected at the time they are laid? Further research may lead to this missing piece of the coelacanth jigsaw.

In 1972 a joint Franco-British-American expedition captured a live specimen and was able to study it for several hours in a small tank. The scientists were surprised to find that its main propulsive force came from a side-to-side sculling movement of the second sets of dorsal and anal fins. The pectoral fins appeared to help stabilize the fish while it moved. The tail fin, which provides the greatest swimming force for most fishes, did not seem to play any part in the coelacanth's movement.

The experts now believe that the coelacanths have survived for

Below: coelacanth specimen caught near the Comoro Islands in the Indian Ocean on June 19, 1960. It was examined by Jacques Millot at the laboratory of comparative anatomy at the National Museum of Natural History in Paris, where he has studied live fish whenever possible. An adult coelacanth ranges from 4 to 5.5 feet long and weighs from 70 to 180 pounds. Alive the fish is a steely blue-gray color, flecked with light spots; after death its color changes rapidly, usually to chocolate brown. The eyes are phosphorescent and they shy away sharply from the light. It survives only a few hours after being brought to the surface. One specimen, though placed in a submerged boat as an aquarium, died within a day through the combined effects of decompression and rise in temperature—they live at depths of 80 to 400 fathoms where the water temperature is about 54°F (the surface waters of the area are usually at about 79°F).

the last 70 million years as an aberrant offshoot of the early fishes, having developed at a tangent from the main line of evolution. They show relationships with sharks, chimaeras, lungfishes, and other primitive fishes, but they are a dying race, last survivors of an independent evolutionary line.

Living fossils can frequently answer questions which arise in the study of rock fossils, and this has been the case with *Latimeria*. Fossil coelacanths show the skull to be hinged at the halfway point, and the brain cavity is very large for the size of the fish. How the brain worked if the skull was hinged had long been a puzzle. Living coelacanths provided the answer: the brain is actually quite small and is located at the rear of the skull behind the hinge. However, it is embedded in fat and cushioned by two large blood ducts, which explained how the large brain cavity of the fossils had completely misled the scientists.

The lungfishes living today are also prehistoric survivors. Lungfishes first made their appearance in the Devonian period nearly 400 million years ago, and reached their peak during the late Paleozoic era, but by the Triassic period 200 million years ago both numbers and species were dwindling. Fossils reveal an almost world-wide distribution, though modern lungfishes are restricted to the rivers and lakes of Africa, Australia, and South America. The living Australian species, *Neoceratodus forsteri*, retains more of the primitive characteristics found in Triassic fossil specimens. The fleshy-lobed, paired fins are flipperlike, and the body is covered with large, overlapping scales. It may reach some 4 feet in length and weigh over 20 pounds, though most specimens caught have been about half this size. All lungfishes have lungs for breathing air, essentially modified from pouched branches in the gut (in most fishes the pouches have been modified into air bladders).

Some authorities have suggested that modern lungfishes show a great resemblance to the newts and salamanders of the class *Amphibia*, and so have proposed that the early lungfishes may have been ancestors of certain four-legged groups. However, detailed research has shown that this is unlikely. Although there are definite similarities, they are probably due to *convergent evolution*—the evolutionary process which produces increasingly similar characteristics between two or more groups of organisms which were initially different and are not closely related. Although a fishlike creature that breathed air certainly crawled out on fleshy, limblike paired fins onto the shore of a steamy jungle over 380 million years ago in the early Devonian period, it was not a lungfish but something that looked a lot like it. But though amphibians are the most primitive class of land-living vertebrates, and the majority of those living today must return to the water to lay their eggs, yet no prehistoric survivors are to be found today. Most of this class have changed greatly.

Reptiles, the class which includes the great dinosaurs, evolved from amphibians and are easily distinguished from their ancestors by their dry, scaly skin which was a great advantage in that it prevented them from drying out. The reptiles were also well adapted to a life spent entirely on land, because their offspring hatch from eggs which develop inside the mother or are laid in or on the ground, rather than in the water.

Above: the living Australian lungfish (*Neoceratodus forsteri*), which has changed little from the Devonian lobe-fins. It probably lives the same sort of life as they did, though it has lost their heavy head armor. The Australian lungfish is the most primitive of the living species and cannot survive out of water for very long. The African (*Protopterus*) and South American (*Lepidosiren*) lungfishes can survive drought by curling up in dug-out chambers in the mud of a drying river bed. Fossil burrows of this type are known from Permian rocks, which proves that ancient lungfishes had the same habit.

Above: tuatara (*Sphenodon punctatum*), an archaic reptile found only in New Zealand. Its vestigial "third eye" is only visible in younger animals. The tuatara lives in burrows by day and is active at night. It feeds on invertebrates and occasionally on the eggs or young of petrels (sea birds which also make burrows). This tuatara is actually sharing a burrow with a petrel.

The resemblance between many modern reptiles and their prehistoric forebears is striking. Try to imagine yourself surrounded by modern living iguanas, frilled lizards, and thorny devils—but all enlarged and 10 to 30 feet long. You would probably think this a good approximation of the Age of Dinosaurs, but in fact the majority of living reptiles are of quite recent evolutionary origin, and few are prehistoric survivors. However, one living species of reptile, the 2-foot-long, lizardlike tuatara, *Sphenodon punctatum* of New Zealand, is the sole survivor of *Rhynchocephalia* ("snout-head"), an order of reptiles which flourished in the Triassic and Jurassic periods at the very beginning of the Age of Reptiles.

When the tuatara was first described in 1831 it was thought to be merely another kind of clumsy lizard. Not until 1867 was it realized that this was a true living fossil. The characteristics which show it to belong to this ancient order are its primitive two-arched skull (more like that of a crocodile in structure than that of a lizard), ribs that have hooklike parts halfway down their length for the attachment of muscles (also found in some birds and fossil reptiles), and the absence of a male organ for copulation (well developed in lizards and snakes). Perhaps the most striking feature of the tuatara is its "third" or pineal eye, situated on the top of its brain just below a hole in its skull. It has the vestiges of a lens and retina but no iris. This "third eye" appears to be an organ for perceiving light, but its actual function has remained a mystery to biologists for many years. Numerous experiments, under conditions as near to nature as possible, have failed to show that it responds to light stimuli. Much painstaking research has as yet revealed no definite function, although several possible ones have been suggested including the regulation of temperature or sexual activity.

The Three-eyed "Snout Head"

The tuatara, although definitely a relic of its ancient order, does not show the typical "snout head" evident in the majority of the extinct members. The fossil rhynchocephalians have a very pronounced snoutlike overhang of the tip of the upper jaw. During the Triassic period the tuatara lived alongside its more specialized herbivorous relatives, some of whom were over 6 feet long, and probably used their toothless parrotlike beaks to crush the hard shells of the seeds of certain ferns. Both tongue and jaws had numerous tiny teeth which helped to rasp any hard and fibrous plants, and which were continually wearing out and being replaced. These beasts had vanished by the end of the Triassic period, but the less specialized *Sphenodon* survived and has continued very little changed for nearly 200 million years.

The tuatara has survived because it has been isolated on a few islands where there are very few predators. Islands are frequently refuges for relic animal forms. The tuatara was found on the main North and South Islands of New Zealand until, with the appearance of man, the life of such faunal relics was greatly altered. To "enrich" the animal life of New Zealand man introduced weasels, pigs, rats, cats, and dogs. By the mid-19th century the tuatara had disappeared from the mainland. Today this living fossil survives only on a few small islands around New Zealand, including those off the northeast coast of North Island and in the Cook Strait. On some of these islands the reptile has enjoyed a population boom due to the New Zealand government's policy of strict protection. It now seems that over 10,000 tuataras are still alive, and although they breed slowly—the eggs take some 15 months to hatch and the animal is not mature until it is 20 years old—if undisturbed they can live for a century.

Islands have also proved safe places for many primitive and ancient birds and mammals. New Zealand, for example, is also the home of the kiwi, a primitive, flightless, chicken-sized, nocturnal bird. Although it now survives under protection, the species did once face extinction due to weasels pursuing it into its breeding burrows. In addition to the small kiwi, prehistoric New Zealand was also the home of the moas, a family of large flightless birds, which could survive because of the absence of land enemies. Some of the 22 species were gigantic, reaching over 9 feet in height. They lived at the same time as the early "pre-Maoris," often called moa-hunters, in whose refuse heaps remains of half-eaten moas have been found. The birds appear to have become extinct shortly after the great immigration of Maori tribes to the islands about 1350 A.D., but legends of the moas persist in Maori tribal law.

From the reptiles evolved the birds some 150 million years ago, and indeed some zoologists regard them as "glorified reptiles." The first bird fossil to come to light, a link between the two groups, was the famous *Archaeopteryx* found in a slate quarry in Bavaria, southern Germany, in 1861. Five more specimens have been found since then. Although not a true bird in the modern sense, neither is it a true reptile. Its lizardlike head, toothed jaws, and long, slender lizardlike tail are reptilian features, but its wing bones, ending in three slender, unfused, clawed fingers, are atypical. It was covered with feathers, which places it without doubt in the line of bird evolution. Of course, compared with the

Below: skeleton of the elephant-footed moa (*Dinornis elephantopus*) of New Zealand, a prehistoric survivor eliminated comparatively recently by man. These great ostrichlike birds were very tall and had a sturdy skeleton, powerful legs, a small skull, and a short, flat beak, which was sometimes curved and sometimes sharp and pointed. The wings and shoulders were undeveloped. Many species of moa lived in New Zealand until a few hundred years ago, where the lack of competition from other land vertebrates provided an unusual opportunity for the flightless birds.

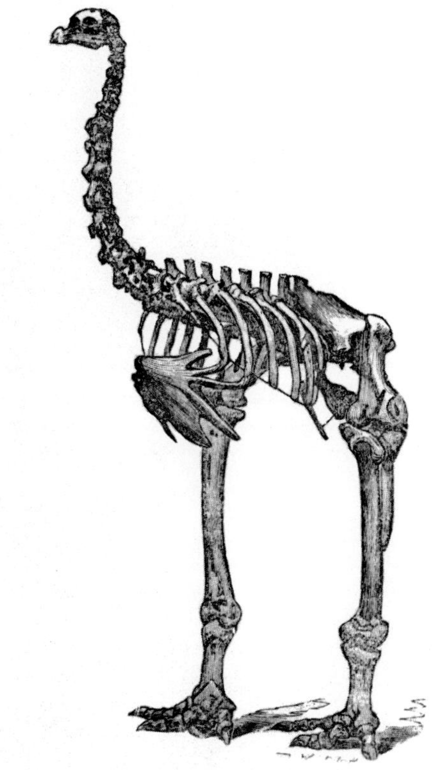

Feathered Survivors?

spectrum of animal life over the last 2000 million years, the birds are of comparatively modern origin. We cannot really call any of them "living fossils," but many avian relics can certainly be found around the world. The majority, however, including the very numerous perching birds (passerines), have evolved quite recently.

One living bird is sometimes called the living *Archaeopteryx*. This is the hoatzin, a strange tropical forest bird of South America which has some archaic avian features. It is slender, somewhat pheasantlike, and reaches about 2 feet in length, and it is characterized by a crest of stiff feathers. It lives in the shrubby trees bordering the tropical streams of northeastern South America. A mated pair will build a platform of sticks in which the female lays two to three eggs. After about a month, an almost featherless chick hatches, showing the species' most interesting primitive peculiarity: two well-developed, functional claws at the tip of each wing. The chicks soon begin to crawl and grip, using all four limbs in a reptilian manner. They use their wing claws to cling to tree branches as they crawl from the nest. Even if the young hoatzin were to fall into the tropical stream below, it would be able to swim well and to crawl ashore using these claws. After two to three weeks the claws disappear, and no vestige of them remains in the adult bird. However, their presence in the chick suggests that they would have been found in a long-vanished and unknown ancestor. An adult bird with wing claws could be one of the many missing links in the evolution of birds from reptiles.

Classification of this survivor is difficult and ornithologists have placed it either with the pheasants and turkeys (the order

Below: hoatzin (*Opisthocomus hoazin*) of northeastern South America on its nest. The adults are crow-sized and are brown with pale streaks on the breast. They have a long tail composed of 10 loosely arranged feathers and an untidy stiff crest on a very small head. The young have claws on the first and second digits of their wings. It is also interesting that the hoatzin has a musky odor reminiscent of a crocodile and that its call sounds more like a reptile's than a bird's. For obvious reasons it is sometimes known as the "reptile-bird."

Galliformes) or with the turacos of Africa, which are commonly classified with the cuckoos in the order Cuculiformes. A recently developed method is the analysis of the eggs of bird species, especially the protein in the egg white. As a result of this technique, the hoatzin is no longer considered a family of its own but a strange type of cuckoo. But the fact remains that the hoatzin is one of those peculiar living relics that has remained relatively unchanged within a restricted range while its close relatives have disappeared. It can provide us with some vague clues to the characteristics which appeared during the development of birds.

The most primitive birds living today include the ostriches of Africa, rheas of South America, and the emus and cassowaries of Australasia. They resemble each other quite closely, and all are large, flightless, ground-dwelling birds. For years they were believed to be related, and fossil evidence shows them to have existed very early in the birds' evolutionary sequence. However, they are now thought to have arisen independently on each of the three separate continents and to be examples of convergent evolution. All have survived due to lack of predators until recent times, when man has become their primary enemy.

Another order of birds which evolved early in avian history is the penguin group. These solemn-looking, flightless, aquatic birds of the southern hemisphere are primitive in some respects, and yet in others they are among the most specialized of birds. Their wings are totally incapable of flight but make powerful flippers for swimming. Some penguin fossil remains show them to have been as large as a man, and to have existed at the dawn of the Tertiary period some 65 million years ago. An interesting hypothesis put forward to explain why these birds are flightless and aquatic suggests that, while at this time mammals were evolving to succeed the dinosaurs, few members of this class reached the isolated Antarctic lands. The ancient penguins may have evolved a ground life (including a loss of the ability to fly) in this empty land, which at the time had a temperate climate. However, when the gradual cooling began soon after, it is possible that, unable to fly away from the increasingly inhospitable land, they took to the oceans and became superbly adapted for an aquatic life.

Ancient animals which are of particularly great interest to man are those that belong to his own group—the class *Mammalia*. The first traces of fossil mammals appear in the late Triassic period 180 million years ago, and primitive mammals continued in existence during the Age of the Dinosaurs. Much can be learned from two curious survivors which show primitive stages probably common at that time. One primitive feature is the habit of laying eggs—a typical reptilian characteristic shown by the duckbill platypus (*Ornithorhynchus anatinus*) and the echidnas or spiny anteaters of the genera *Tachyglossus* and *Zaglossus*, which are found in Australia, Tasmania, and New Guinea. Surprise and disbelief followed the discoveries, in 1884, of two warm-blooded, hairy animals which were found to lay eggs. Professor Wilhelm Haacke found that echidnas laid eggs almost at the same time as the British zoologist H. W. Caldwell made a similar discovery in Queensland, Australia in regard to the platypus. Before this time the actual existence of the platypus had been a matter of con-

Below: jaw of a *Megazostrodon*, one of the earliest known mammals. Only the size of a shrew, it evolved in the late Triassic period with a hairy covering to keep heat loss from becoming unmanageable. It may have lived alongside the last of the dinosaurs

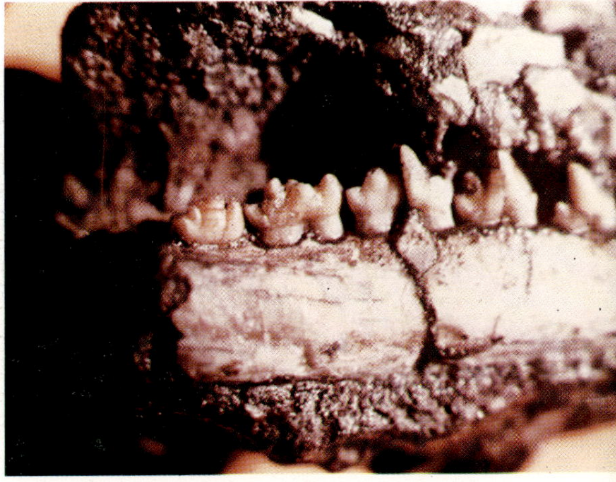

Right: duckbill platypus (*Ornithorhynchus anatinus*) surfacing to breathe. A semiaquatic creature, the platypus has webbed feet for swimming and a flat, ducklike beak for scooping up worms and other small creatures from the river bed. It also uses its beak to dig the burrow in which it lays its eggs. The duckbill platypus grows to about 20 inches in length. Monotremes are the most primitive mammals and the most closely related to reptiles.

troversy, and when the first complete skin arrived at the British Museum in 1798 its authenticity was doubted. And why not, when the beaverlike pelt had a beaver's tail at one end and at the other the dried bill of a duck. At this time there were many animal frauds such as stuffed mermaids consisting of a withered monkey's head carefully attached to the rear half of a stuffed fish, or gorgeous "new" specimens of bird-of-paradise created by imaginative Asian taxidermists. However, four years later with the arrival of several complete carcasses, the platypus was accepted as a real creature. It shares several features with its reptilian ancestors, including a single passage for the elimination of excrement, urine, semen, or eggs (most higher mammals have two passages). *Monotreme*, the biological name given to the platypus and echidna, actually means "one-holer."

Laying eggs with tough, leathery shells is a feature for which monotremes are noted. A bony shoulder girdle, poor temperature control, and the ability to fast for long periods are also reptilian characteristics. The platypus sits or rolls onto its back and keeps its eggs on its belly for 13 or 14 days, incubating them like a bird. The female echidna develops a small pouch at the beginning of the mating season, and after mating she later transfers a single egg into it. Neither platypus nor echidna females have nipples, but their mammary glands discharge milk which is sucked off their hairs by the young.

Early mammalian creatures no doubt laid eggs, but the first mammals certainly did not look like these survivors. The platypus is a highly specialized creature which leads a specialized life. Toothless as an adult (although the young have milk teeth), it has become adapted for an aquatic life—it lives mainly in streams and has good digging feet for nesting in burrows on the banks. The spiny anteater has become just as specialized, being protected by a stout, spiny covering comparable to that of the hedgehog and equipped for its ant-eating habits with a long, slim snout (it is toothless from birth), sticky, protrusile tongue, and powerful digging feet for ripping up ant and termite nests. But for all this, these curious creatures are still links with the mammals' reptilian ancestors. They have survived not only due to isolation but to a mode of life in which competition is minimal.

Australia is also the home of other comparatively unprogressive mammals—the *marsupials*, which have pouches for carrying their young. The babies are comparatively undeveloped and look like tiny embryos at birth. These primitive mammals are also found on the South American continent. But the flaw in the process of bearing young alive is that if they are too tiny and helpless they will not survive—there must be a means whereby the young get sufficient food before they are born. The marsupial pouch was fairly effective, but the other main group of mammals managed to find a better way. They evolved a *placenta* from one of the membranes that surrounded the developing egg of their reptile ancestors. The placenta is the lifeline between mother and offspring, transporting food and oxygen from mother to embryo and getting rid of waste materials in the opposite direction. The embryo could thus grow to a far higher state of development than had previously been possible. The higher mammals evolved the placenta very early on, to which their success was due.

Fossil beds have been found to show that in the late Cretaceous period while dinosaurs ruled the earth the two living groups of mammals coexisted with them—the marsupials and the placentals. There are two groups of marsupials, of which one, the opossum, is the more primitive and today is found only in North and South America. The living opossum species, such as the common Virginian opossum, are typical marsupials, in habits and structure very similar to the Mesozoic mammals. When the early marsupials began giving birth to live young the females had not yet evolved a satisfactory way to feed the developing baby inside their body, the eggs containing comparatively little yolk. As a result the baby was born at a very tiny and immature stage. But it could continue its development inside a pouch which evolved to grow during the breeding season. Though like an embryo in appearance, the tiny baby had strong forelimbs to help it crawl from the *cloaca*, the cavity into which it was born, up the mother's body into its "incubator." Here it attached itself to its mother's nipple and—nourished, warmed, and protected—completed its development over the next several weeks or months.

During the Age of Dinosaurs there were numerous opossum species, very like the surviving species but much smaller in size. As the Age of Mammals was dawning some 65 million years ago they were widespread over the earth, but in most places they made little progress and were eventually supplanted by higher mammals.

Marsupials in South America and Australia survived because those continents had drifted away from the other major continents before the higher mammals reached them. (Continental drift is discussed in Chapter 2.) Some 180 million years ago at the

Primitive Mammals

Below: spiny anteater or echidna (*Tachyglossus aculeatus*), the second monotreme to be discovered. Unlike the platypus it is a land animal, and its habitat is the sandy and rocky parts of Australia, Tasmania, and New Guinea. The feet are not webbed but have pads on the soles of their feet. It has strong claws for burrowing, a rudimentary tail, and a convoluted brain. The sexes are externally alike. The one or two eggs laid by the female are placed in a rudimentary marsupial pouch which develops to coincide with laying, and the mammary secretion is discharged into this pouch. Its gait is ungainly because of its claws, but it is capable of moving very rapidly. It can burrow out of reach if frightened, or it can roll up like a hedgehog, offering nothing but spines to an aggressor.

beginning of the Jurassic period Australia was probably still attached to Antarctica but had separated from Africa, and soon after it severed its last connections with South America. At this time some primitive egg-layers and marsupials were aboard the Australian "ark" and henceforward these evolved in isolation. About the same time South America separated from Africa and was drifting away from it, connected only at the Amazon-West Africa area. Water completely separated Africa and South America by the middle Cretaceous period, 100 million years ago. The South American continent was probably not connected to North America either at the dawn of the Tertiary period some 65 million years ago, so that the fauna could evolve or survive in comparative isolation for many millions of years. Many strange and curious herbivores evolved; some opossums evolved into large flesh-eaters that paralleled the development of wolves and cats on other continents. However, about 2 million years ago during the Pliocene epoch connections were reestablished with North America, and soon several placental flesh-eaters had moved south. In time they had wiped out the carnivorous marsupials, but the small, *omnivorous* (both plant- and meat-eating) opossums survived.

Indeed, opossum ancestors seem to lie at the base of the Australian marsupial's evolutionary tree, and in the absence of placental mammals until man arrived the group radiated into an interesting range of fauna. Herbivorous forms evolved, including

Below: opossum (of the family Didelphidae) with young in its pouch. Sedentary animals, they rarely wander from their locality. Males seem to be more numerous than females, and the adults live in isolated pairs. In wooded districts the opossum is arboreal, sleeping in tree hollows, but in the open plains it lives in burrows, sometimes enlarging burrows dug by other animals until they are about 5 feet deep. Sometimes it lives with other animals such as the armadillo. In some places its fur is much prized, particularly in winter when it is at its thickest. The opossum is extremely fertile and thus does not appear in any danger of extinction, although it has been extensively hunted for a century.

The Shrews as Survivors

Left: the Tasmanian Devil (*Sarcophilus harrisii*), which has an exaggerated reputation for savagery. Once widespread in Australia, it is now confined to Tasmania. It is stocky and powerful with a large head and strong jaws and teeth; it rarely exceeds 3 feet in length. Though a land animal, it haunts beaches and the banks of rivers in search of food, which is made up of bandicoots and other small herbivores.

pouched "flying squirrels," koalas or "native bears," the hopping kangaroos, and the wombat, a woodchucklike animal. The sparse fossils, dating from some 2 million years ago, have revealed that giant kangaroos and rhinoceros-sized wombats once existed.

On the other hand, primitive opossum types also gave rise to carnivorous marsupials, some small such as the native arboreal cats and other larger ones including the Tasmanian devil—a powerful wolverinelike flesh-eater—and the thylacine or Tasmanian wolf, wolflike in appearance and behavior but nevertheless a pouched mammal.

Ancestral placental mammals, on the other hand, were rather general in their food habits, but they primarily ate small insects. A few forms still exist to this day which have retained such feeding habits, and these seem on the whole to have departed the least from their primitive placental ancestors. These are the *insectivores*, the shrews. They show more or less what our ancient Mesozoic ancestors must have been like. These small, mouselike creatures with long, sharp snouts led secretive, hidden lives, and it is still a mystery why they remained so small while the dinosaurs ruled, and why only on the downfall of these giants did they grow larger, more powerful, and begin to radiate into the great array of mammals found living today, from rats to horses, otters to whales, and tree shrews to man. Whatever the reasons, the living shrews give some insight into how certain animal stock lived millions of years before the evolution of primitive man.

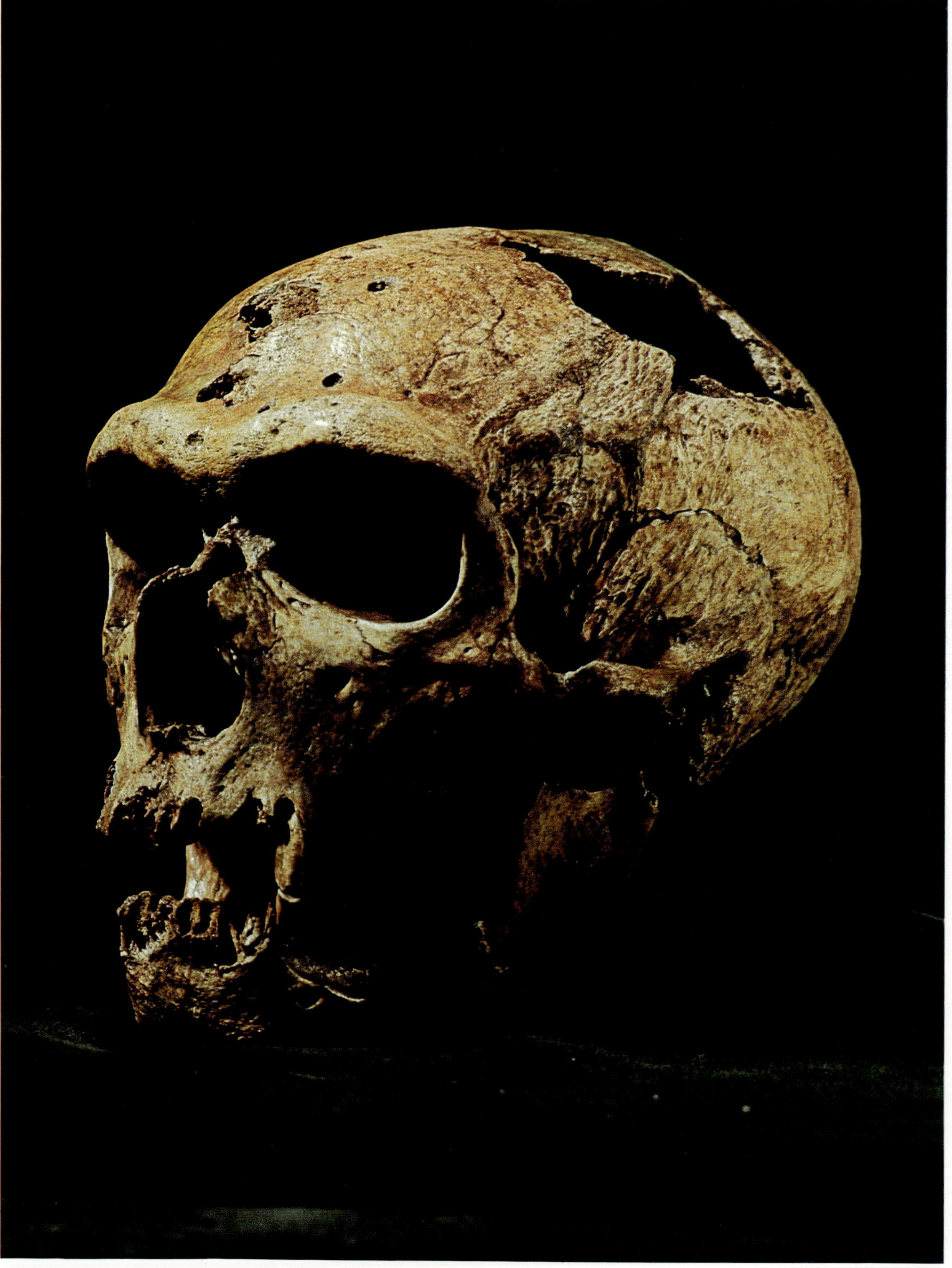

Chapter 6
Descent from Ape or Adam?

The subject Charles Darwin deliberately avoided in his great work *On the Origin of Species* was a contentious and potentially explosive one—the origin and evolution of man himself. Later he allowed himself to suggest that man was descended from the apes, but it has not been until comparatively recently that fossil finds have corroborated his suppositions. Archaeologists working in East Africa have been searching for the "missing link" between modern *Homo sapiens* and the gorilla and chimp family, but new finds constantly undermine their most careful theories. From the hoax of "Piltdown man" to the lineage of *Australopithecus*, the subject of man's origin is proving as fresh and challenging as ever—and who knows what the next decade may reveal?

Before Charles Darwin published his *On the Origin of Species* in 1859, he wrote to Alfred Russel Wallace, the naturalist and explorer who, independently of Darwin, had formulated a theory of natural selection while living in the Malay archipelago (now Indonesia). Darwin informed Wallace that he would avoid the whole subject of the origin of man because so many prejudices surrounded it, although he admitted it was extremely interesting and of prime importance to naturalists. So in his book Darwin confined himself to the modest comment on natural selection that "much light will be thrown on the origin of Man and his history." His hostile religious opponents wanted to know what evidence linked man with the gorilla, chimpanzee, and orang-utan, but Darwin did not enter the public debate until 1871 when he published *The Descent of Man and Selection in Relation to Sex*. He pointed out comparisons between man and apes in physical form, physiology, susceptibility to disease, instincts, emotions, and sociality to support his thesis that modern man was descended from the apes. What Darwin did not have, which makes more amazing the accuracy of his surmise, was the evidence of fossil bones of man's predecessors. Darwin also guessed that Africa was where man had originated, simply because the gorillas and chimpanzees, probably man's nearest relatives, are found only on this continent.

Opposite: Neanderthal skull found in a cave at La Chapelle-aux-Saints in southwest France in 1908. One of France's leading anthropologists, Pierre Marcellin Boule (director of the Natural History Museum in Paris), was called to the cave by three members of a monastery who had discovered bones while digging for tools of the Mousterian period. Boule himself then uncovered this Neanderthal man in first-class condition, buried about 2 feet down. He had obviously been given a formal grave. One observer described the find as follows: "The head of the dead man lay on a bed of stones, looking westward, with his right hand tucked under his head as if in sleep. The legs were drawn up against his body, and over him lay the bones of reindeer and ox, probably originally intended to provide him with food." A hole in his skull seemed to indicate a cult cannibalism similar to that of the hunting nomads of the later Stone Age.

Above: cartoon from the *Hornet* of March 22, 1871 showing Charles Darwin as a grotesque ape. Although Darwin himself never suggested that man had descended from apes, the idea called forth a great deal of controversy and vicious criticism of him and his adherents by people who preferred to believe the biblical version.

But, although Darwin did not know it, fossil fragments of early man had been discovered in the summer of 1856 in the Neander valley, a steep-sided gorge a few miles from the town of Düsseldorf, West Germany. Workmen blasting there opened a small cave, inside which were many ancient bones. Totally ignorant of the meaning of their find, they had smashed many bones by the time scientists were called to the scene. All that remained were a skull cap and a few fragments from the rest of the skeleton.

The skull bones were unusually thick, the forehead was low, and the projecting eyebrow ridges were heavy. The overall picture contrasted strongly with the skull of modern man, *Homo sapiens*, which is domed and characterized by a high forehead, thin bones, and poorly developed eyebrow ridges.

At the time of the discovery, however, the bones were rejected as being unrelated to human evolution. Professor F. Mayer of Bonn University, rejecting any sort of evolution theory, ingeniously supplied a different but acceptable answer: the bones must have belonged to a Mongolian cossack in pursuit of Napoleon's army after his disastrous invasion of Russia. The cossack had probably rested in the cave and died there—no doubt from fatigue as well as an advanced case of rickets, evidence for which were the prominent brow ridges of a furrowed forehead, the effect of the pain of the disease. Here his body remained until dug out by the workmen.

The majority of interested scientists were similarly skeptical, and the consensus was that the bones were somehow of recent or contemporary origin. The unusual characteristics were thought to be the result of disease or abnormality. One suggestion was that the man had been a criminal! Although we may now find these ideas amusing, it is true that some *Homo sapiens* do have similar skull peculiarities due to disease, abnormal growth, or malformation at birth. However, as in many branches of science, scientists differed among themselves. Darwin's friend, the British biologist Thomas Henry Huxley, argued in his book *Man's Place in Nature* (published in 1863) that the Neander fossil finds were an early human type although not in a direct ancestral line to modern man.

Interest was sparked off, and scientists and laymen alike began to search for similar finds. A previous discovery in Gibraltar in 1846, of a skull with a skull cap, massive jaw, broad nose, and very large eye sockets, was considered by some as a type of primitive man, and these fossil forms soon became known collectively as "Neanderthal Man," *Homo neanderthalensis*.

In 1886 two more skeletons of Neanderthal man were found at Spy, a small village near Namur in Belgium. They were better preserved than the others and had the same characteristic features. Primitive faces of early Neanderthal man were built up from the available information, and the result was a flat-headed human with large eye sockets overhung by a bony shelf, strong muscular jaws and a receding chin, and a scanty forehead with much of the brain situated at the back of the head. *Homo neanderthalensis* was also confirmed to be a man rather than an ape because stone implements and mammalian bones (indicating their use as food) were found at Spy along with the human bones. The

The Discovery of Neanderthal Man

animal bones were those of a woolly rhinoceros and a mammoth, both of which roamed the area during the last period of the Ice Age over 10,000 years ago.

Neanderthal man was finally accepted as a true human type—a *hominid*—when an entire skeleton was found in a cave at La Chapelle-aux-Saints in southwest France in 1908. The term *hominid* is used to describe a family containing man and his close relatives as distinct from the anthropoid ape family, the *Pongidae*. The main differences between man and the pongids relate to feeding and locomotion. Man stands upright and is bipedal—that is, he walks on two legs. All apes walk on four legs, though they are capable of rising up onto their hind legs to cover short distances, when they balance themselves with their long arms. Man's bipedal stance has influenced the shape of his hip bones, thigh bones, and foot bones, and has given him his characteristic S-curved backbone.

Dental characteristics imply different feeding methods and diet. The short human face, compared with the jutting face and jaws of the apes, is due to the teeth. Human teeth are arranged in a continuous arcade, all on the same level. In apes, on the other hand, the cheek teeth of one side of the jaw are parallel to those of the other side. They terminate in large, projecting canines, with a small gap between them and the smaller sharp biting incisors at the front of the jaws. The gap is so that the canines fit into the mouth when the jaws are closed.

Many pieces of information are still missing, though, and the classification of direct ancestors of *Homo sapiens* is still somewhat confused and unsettled. The episode of the so-called "Piltdown man" was a mystery that took over 40 years to solve. It illustrates the unfortunate weakness of some scientists in promoting certain discoveries which seem to be in line with their own area of research.

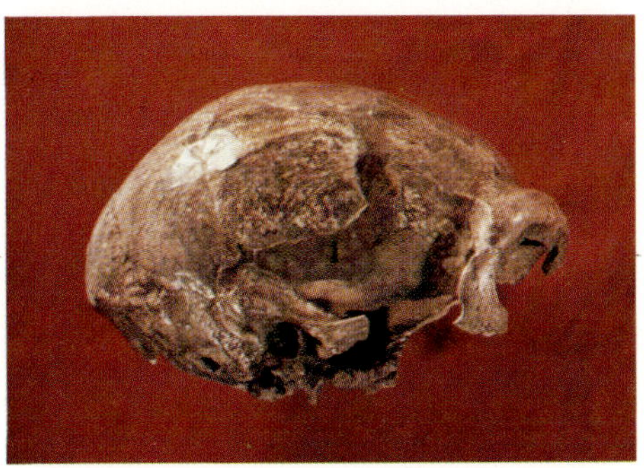

Above: one of the skulls found at Spy, near Namur in Belgium, in 1886. Three Belgian geologists, Lohest, de Puydt, and Fraipont, found the two adult male skeletons about 12 feet down in a layer of earth previously untouched.

Below: reconstruction of the life of Neanderthal cavemen at Le Moustier cavern in the Dordogne region of France. Bones of large animals found there suggest that Neanderthal hunters used intelligent cooperation. The driven game, killed and butchered, would then be shared out among them.

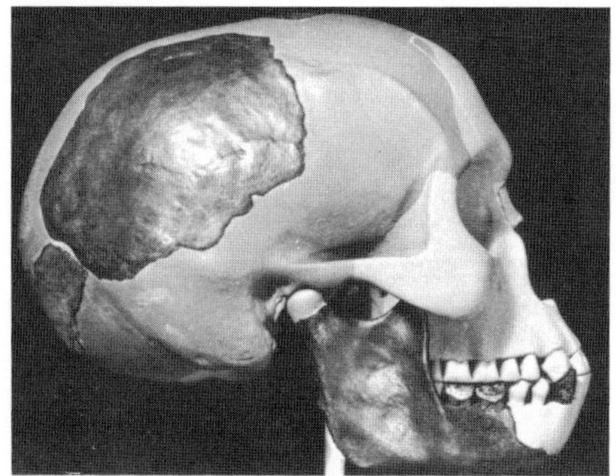

Above: the skull of Piltdown man, the white areas of which are the "missing" parts as reconstructed. In 1908 Charles Dawson, a lawyer and a quiet, diffident man, produced a brown cranium found in a gravel ditch near Piltdown, Sussex. He had apparently been given it by someone who worked as a streetdigger. He found more bones himself over the next few years, and in 1912 he got in touch with Sir Arthur Smith Woodward of the British Museum. Smith Woodward himself found a jaw and a few teeth on the site. When he reconstructed the skull the result was "Piltdown man."

Below: Philip Tobias, a respected anatomist and brain specialist, holding the Taung skull found by Raymond Dart in 1925.

Charles Dawson, an amateur naturalist and archaeologist, discovered skull remains in a gravel pit on Piltdown Common in Sussex, England, in 1908 and 1911. Later, together with Sir Arthur Smith Woodward, a noted paleontologist of the time, further remains including part of a lower jaw were found. Smith Woodward reconstructed the skull and named it *Eoanthropus dawsoni*—"dawn man"—as it revealed an apparently very early form of man with a hominid cranium but a very early apelike jaw. Could this be the "missing link" between man and his allies, the gorillas and chimps? Some paleontologists were eager to accept the validity of the find, many were doubtful while others rejected it as a fraud.

The fossil "Piltdown man" mystery was not finally solved until 1953 when chemical tests conclusively demonstrated that, though the skull fragments appeared to be genuine, the jaw was that of an orang-utan which had been deliberately faked to give the impression of a very old fossil form, and that the teeth had even been ground down to the flatness expected in a human. The mystery partially remains, however, since those involved in the hoax have died and it will probably never be known who thought up and executed the fraud, or why.

Probably as a result of the skepticism surrounding Piltdown man, the authentic finds of Raymond Dart, a professor of anatomy at the University of Witwatersrand in Johannesburg, South Africa, were received with wide disbelief. Dart made his outstanding discovery, of a hominid skull, in 1925 near Taung, Bechuanaland (now Botswana). It was that of a child about five years old, with an almost complete face, a lower jaw with teeth, and an impression of the right side of the brain. Dart believed he had found a primitive human form that warranted a new genus, *Australopithecus* ("southern ape"). The main objections to Dart's find were the youth of the skull and the fact that it had come from such a distant and "barbaric" land. Dart was not dissuaded from continuing his work but for the next 10 years he did not find any further fossil human bones at Taung. In 1936 however, the South African Dr. Robert Broom, one of the few who had taken an interest in Dart's work and believed in his findings, heard of new discoveries unearthed during mining operations at Sterkfontein, 30 miles outside Johannesburg. For the next two years Broom and his assistants found fragments of adult australopithicine skulls, jaws, teeth, and other skeletal parts, as well as the bones of animals on which the "ape-men" fed. It was not until after World War II that Professor Dart excavated an almost perfect skull of the *Australopithecus* type and later, in August 1948, a nearly perfect pelvis. This gave evidence for the erect posture of the hominid. Gradually *Australopithecus* lost its "skeleton in the cupboard" status, when paleontologists did not recognize it, and became the center of scientific discussion on the origin of man.

Other evidence of man's ancestors had meanwhile come from other sites around the world. In the 1920s teeth, a nearly complete skull cap, and various bone fragments had been discovered in China in a cave at Chou-k'ou-tien south of Peking, thus establishing another early human form which became known as Peking man. Other discoveries were made near Peking and in Java, and by the late 1940s it was possible to speak of a single

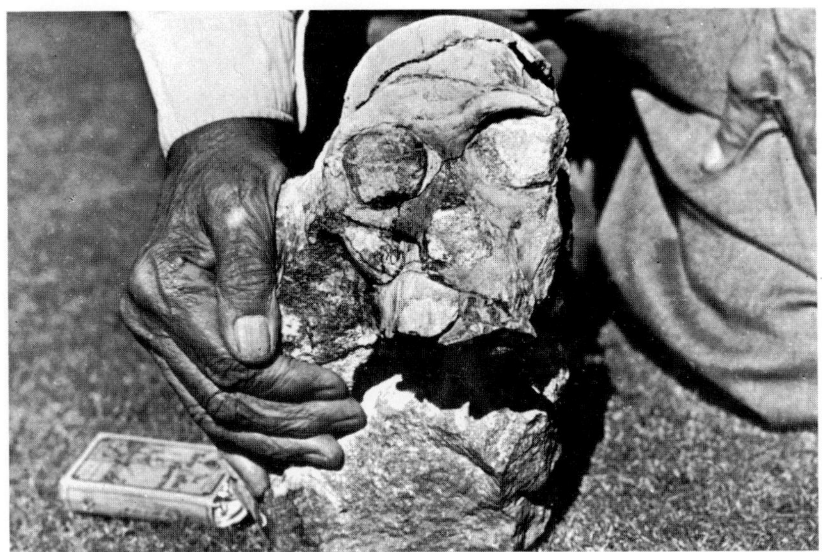

More Finds–More Problems!

Left: anthropologist Robert Broom's discovery of the "Mrs. Ples" skull at Sterkfontein in 1947. The female skeleton as reconstructed was named as an australopithecine *Plesianthropus transvaalensis*, but since then "Mrs. Ples" has been reclassified *Australopithecus africanus* along with the Taung child.

genus *Pithecanthropus*, of which one species, *P. erectus*, was typical of East Indian finds and another, *P. pekinensis*, was typical of the Chinese.

After World War II new scientific techniques were invented, namely the carbon-14 and potassium-argon dating methods, which accurately determine not only the age of the strata in which fossilized bones are found, but sometimes the age of the bones themselves. The Java and Peking men, now classified as *Homo erectus*, were dated and found to be nearly half a million years old.

Professor Louis S. B. Leakey, son of a British missionary in Kenya, and his wife Mary, both distinguished 20th-century archaeologists, were prominent in the discovery of evidence of man and his roots. Their first exciting find was in the summer of 1959 in the Olduvai Gorge, a torrent-eroded cleft of the Rift Valley in northern Tanzania. The gorge is 30 miles long and some 300 feet deep. In late Tertiary times around 10 million years ago it was a lakeshore region, but during the course of time the waters receded and erosion has revealed a beautiful succession of strata rich in fossils.

The Leakeys' find consisted of a highly fragmented skull as well as primitive stone tools and associated dwelling sites. The skull was in more than 400 pieces, so it is no wonder that it took nearly a year to put together the jigsaw. Leakey named it *Zinjanthropus boisei* ("East African man"), but recently it has been put in the genus *Australopithecus*. It was then the oldest human fossil ever found—carbon radioactive dating, based on potassium-argon content, showed it to be 1,750,000 years old.

Within two years the gorge had yielded a second hominid, with a bigger brain but smaller frame than *boisei*. *Boisei*'s brain had a capacity of about 32.3 cubic inches, while the new find had about 40 to 42.5, and a modern human's is about 85.5. This new skull was considered human because of the larger brain size and hominid teeth. In 1964 it was given the specific name of *Homo habilis*, meaning "handy man," because it was supposed that he was one of the first toolmakers. Since that time more fossils have been discovered, many of which are important to

Below: reconstruction of Java man, *Pithecanthropus erectus*, later reclassified as *Homo erectus erectus* ("upright-standing man"). Dubois and von Koenigswald found fossils of this type of man in Java at Sangiran and Trinil, two spots on the Solo river about 40 miles apart.

Above: Louis and Mary Leakey comparing the skull of an australopithecine (left) with that of *Zinjanthropus boisei*, which they found at Olduvai in 1959. They discovered the first skull, of a young male hominid, in among the broken bones of predators.

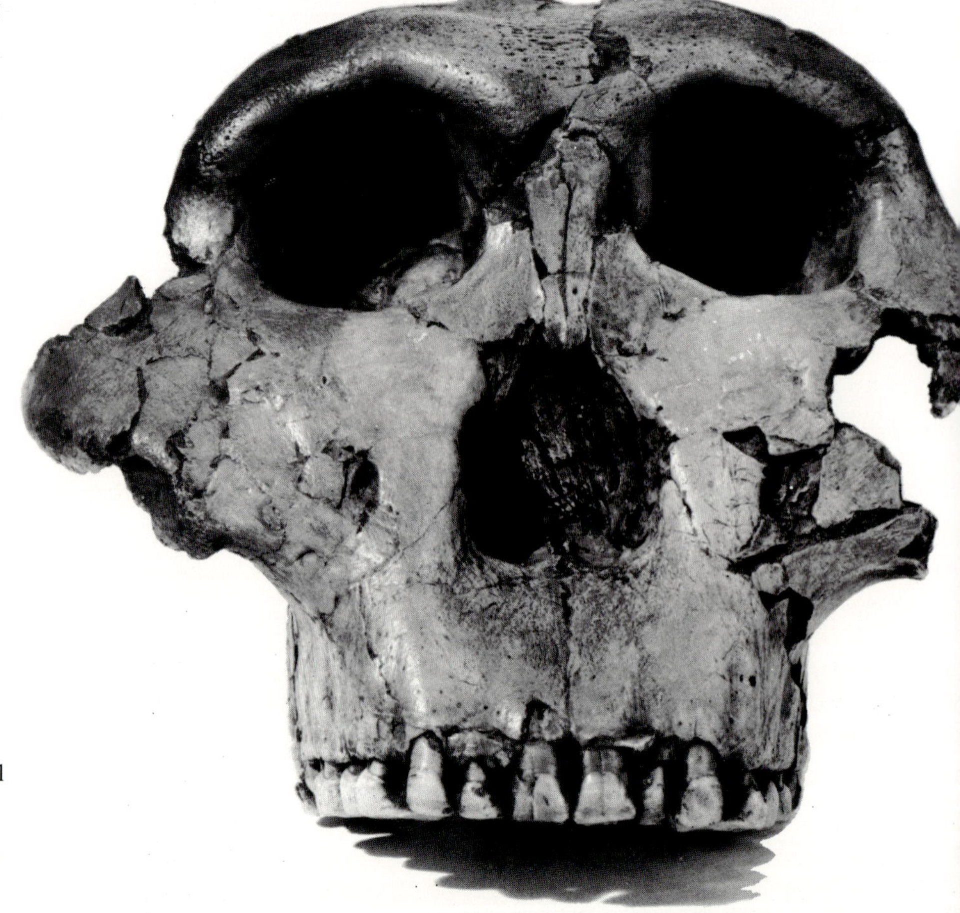

Right: *Zinjanthropus boisei*, an example of the more robust type of *Australopithecus*, characterized by a heavier facial skeleton. The front teeth are small and ill-fitted for meat-eating. Large nutcrackerlike back teeth necessitate a large jaw, a strong muscular system for chewing, and a sagittal crest for the attachment of certain muscles. All these factors point to the probability that *Australopithecus* (*Zinjanthropus*) *boisei* was a specialized plant eater who required no tools for killing game.

man's evolutionary line. There are now established representatives of *Australopithecus boisei*, *A. africanus*, and *Homo habilis*.

What is interesting about these three forms is that they all lived at more or less the same time, from about 3.5 to 1.5 million years ago. Today most anthropologists believe that these types evolved from early apes who lived on earth as much as 25 million years ago. Another theory is that this ancient ancestor probably gave rise to the gorilla and chimpanzee as well. *Dryopithecus africanus* ("woodland ape") is thought to be the stock from which all modern apes have evolved. Another Asian species, *Gigantopithecus*, seems to be the ancestor of the large terrestrial apes of that region which have since become extinct. A third genus, the monkey-sized *Ramapithecus*, could prove the most exciting fossil primate now being investigated, because it is very old (over 10 million years) and may well be on a direct line to the hominids, a true ancestor of man.

Apes, Men and Plate Tectonics

A year after Louis Leakey's death in 1972 the Dutchman Adrian Kortlandt put forward some interesting theories about the African Rift Valley ape and its relation to hominid history. He asked how the course of evolution changed to produce an omnivorous primate (proved by the teeth) who then left an arboreal life to walk erect (shown by the structure of the pelvis) and eventually to use tools (shown by the hands, site tools, and animal remains). Kortlandt used the theory of plate tectonics (continental drift) to clarify the evolution of apes and man: he felt that in and around Africa, where many primates appear to have evolved originally, small changes due to the movement of continental plates had far-reaching effects on primate species. For instance, some 25 million years ago when Arabia may still have been connected to Africa, the ancestral apes of Africa spread outward to Europe and Asia. These apes were monkey-like in habits, living in trees, but they had apelike teeth.

Below: *Homo habilis* cranial fragment, found by Louis Leakey at Olduvai in 1963. Believed to be 1.75 million years old, this direct ancestor of modern man was manually skilled enough to fashion and use tools. As one venerable anthropologist remarked, "The gulf between animal and man is not great—neither is that between genera Australopithecus and Homo."

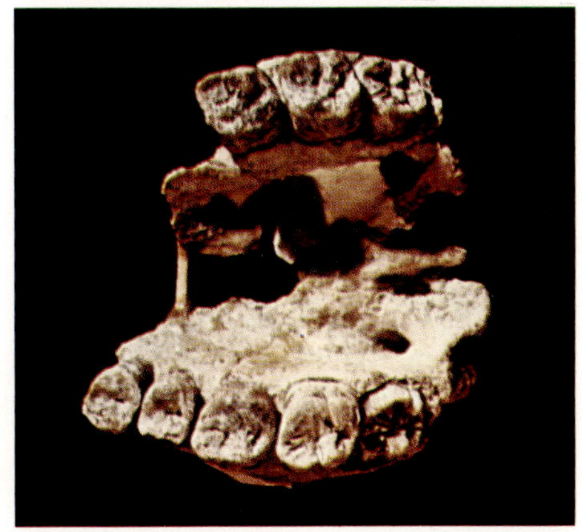

Later the Red Sea began to form as a fault pushed forth sediment and formed an ocean floor between the two continents. The Suez area was also flooded along most of its length. The primitive apes, now cut off from others of their kind, diverged to evolve in their own way. The orang-utan, a species which has never come down from the trees to lead a semi-terrestrial life, was the result in Asia. In Europe and Africa the early apes evolved adaptations which meant that they spent a good deal of time on the ground (this is shown by fossils of their feet). Kortlandt believes that European apes became extinct because they could not compete with the huge bears that dominated the available food of that area. But in Africa the majority of apes were safe, as the large bears did not try to cross the immense stretch of sands of the Sahara desert.

Then 20 million years ago the Rift Valley began to form, followed by a flooding of the new valley floor. Fifteen million years ago the result was a more or less continuous chain of rivers and lakes running from Lake Tanganyika in East Africa to the Nile in the north and from Lake Malawi to the Zambesi river in the south. This split the ape population of the area, since apes do not like to swim and will not enter water. Even if the apes had attempted the crossing, a large concentration of crocodiles guarded the total length of their water habitat. Kortlandt suggests that the

Recent Finds in East Africa

apes living to the west of the rift evolved into two types—the gorillas, who evolved adaptations for dense forest living, and the chimpanzees, who adapted to more open forested country.

The population in East Africa was more or less sealed off, and Kortlandt believes it possible that these ancestral apes eventually evolved into man. This region, comprising Tanzania, Kenya, and southern Ethiopia, became drier and less densely forested over the next 15 million years, so that the animal which evolved over this period adapted to a ground life: he became bipedal and developed grasping hands, a bigger brain, and excellent eyesight. These physical characteristics were accompanied by new social and hunting behavior, and the animals developed a high level of intelligence.

Until the 1970s the view held by most anthropologists was that man had been walking the earth for just one to two million years. During the present decade, however, several exciting finds have shown that man has probably been around for much longer—perhaps as long as 14 million years. In 1972 an international team working in the Afar region of Ethiopia found tools dating back 2.6 million years. An even more dramatic find was made two years later when they discovered, not far from the tools, a large part of the skeleton of a specimen of *Australopithecus*, a 20-year-old female, found in a three-million-year-old sediment strata. Named Lucy after the girl in the popular Beatles' song "Lucy in the Sky with Diamonds," she had definitely walked erect and was just over 3 feet tall with a brain about a third the size of modern man's.

Other researchers had found remains of *Ramapithecus* in several parts of the world, including East Africa, central Europe, the Middle East, and Pakistan. *Ramapithecus* had been known for many years (the first specimens were discovered in 1934), but it was not until 1969 that it was suggested that this small primate could be on the direct line leading to the hominids. Modern dating techniques have been used to establish this genus as about 10 to 14 million years old.

Below: Lucy, a 20-year-old female *Australopithecus* found in the Afar region of Ethiopia, at the northern end of the East African rift valley, in 1974. The bones are remarkable for their small size, although there is no evidence of immaturity.

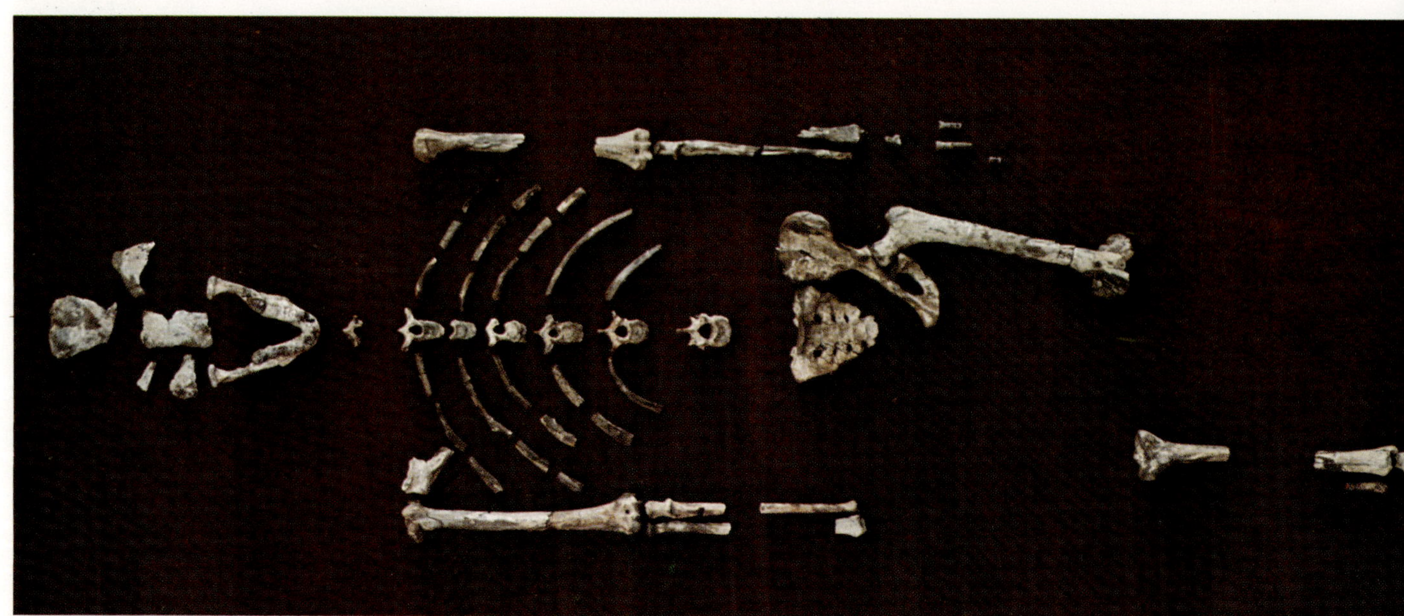

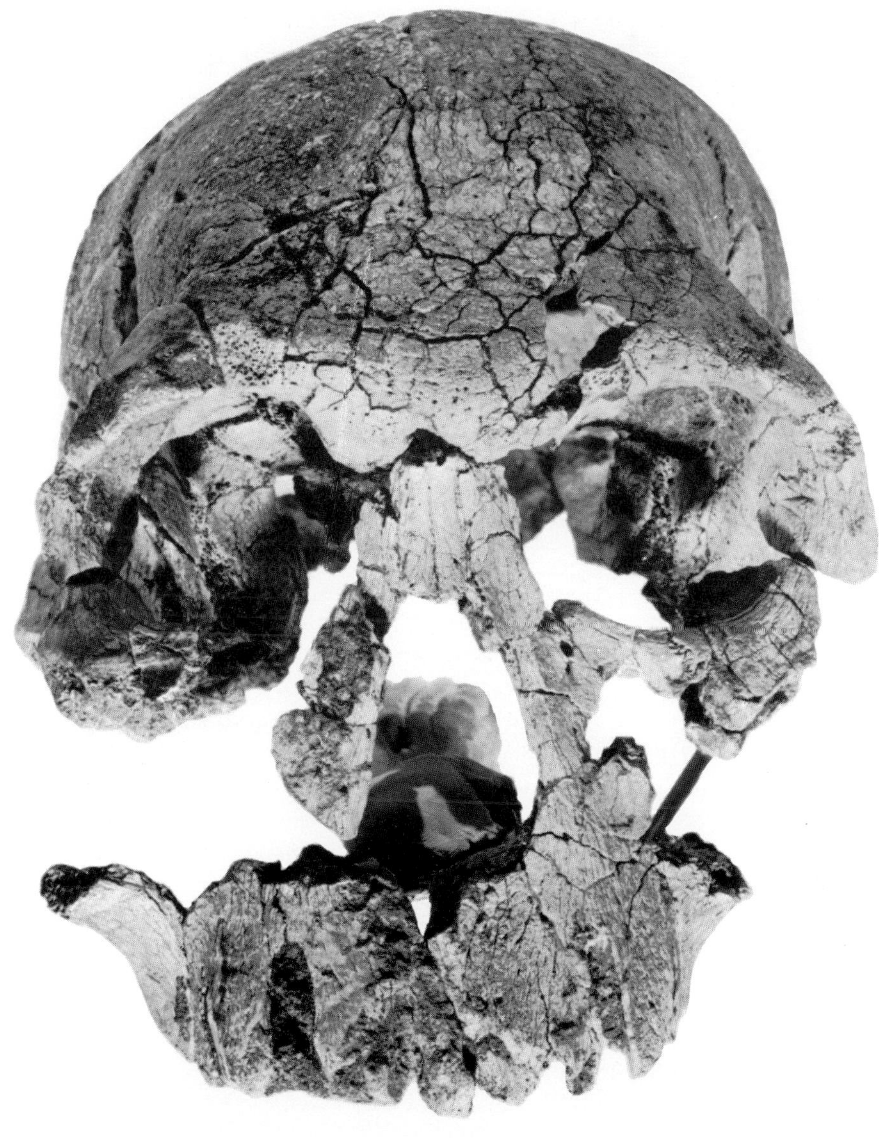

Left: frontal view of the skull named 1470, which Richard Leakey's expedition uncovered on the shores of Lake Rudolf. Leakey himself thinks this is a "prototype of *Homo erectus*." In any case, it has been found to be over 2.5 million years old. Other fragments of jawbones and teeth found by British, French, and North American scientists in this area of East Africa have been classified as nearly 5 million years old.

But it is Louis Leakey's son Richard, born in 1944, who has provided the most exciting recent find. In 1963, after three years working in his own safari business in East Africa, he spotted from his aircraft interesting sediment beds near Lake Natron in northern Tanzania. He decided to explore the area, and he returned with a fragment of *Australopithecus robustus*. In 1967 he joined his father's expedition to the Omo Valley in Ethiopia, and the next year received funds of his own from the National Geographic Society to investigate a site near Lake Rudolf in the Turkana region of northwest Kenya. Shortly after his father's death Leakey and his team discovered a skull, since named "1470" (its catalog number), and later another skull labeled "1590," each of which proved to be specimens over two million years old of *Homo habilis*—as a matter of fact, they were nearly three million years old! This meant *H. habilis* must have coexisted with *Australopithecus*, which suggested that *Australopithecus* was not on the direct line leading to modern man.

Anthropologists have now been left in an interesting quandary. Richard Leakey and others think that human lineage probably

Unanswered Questions

Right: comparing the skulls of a prehistoric and a modern lemur (*Adapis*). The lemur evolved in the late Eocene period and developed many of the specializations that characterize later primates, such as reduced snouts, forward-pointing eyes protected from behind by a bar of bone and providing stereoscopic vision, relatively larger brain size, nails instead of claws, and grasping hands and feet. The lemur is a semiapelike forebear of Old and New World monkeys as well as the ancestor of anthropoid apes, and therefore of man.

began over 13 million years ago with a population of *Ramapithecus* which diverged to form two species of *Australopithecus*, *A. robustus* and *A. africanus*. It also developed along a separate line from which arose *Homo habilis* and eventually *H. erectus* (about 1.6 million years ago). *Homo sapiens* evolved from *H. erectus* about 100,000 years ago. Man as we know him today did not emerge until about 25,000 years ago.

There are still many gaps to be filled in the anthropological history of man. No fossils have yet been found dating from the 10 million years between *Ramapithecus* 13 million years ago and *Australopithecus* and *Homo habilis*. Will the next decade of the 20th century produce new evidence and bring forth new theories of man's evolution?

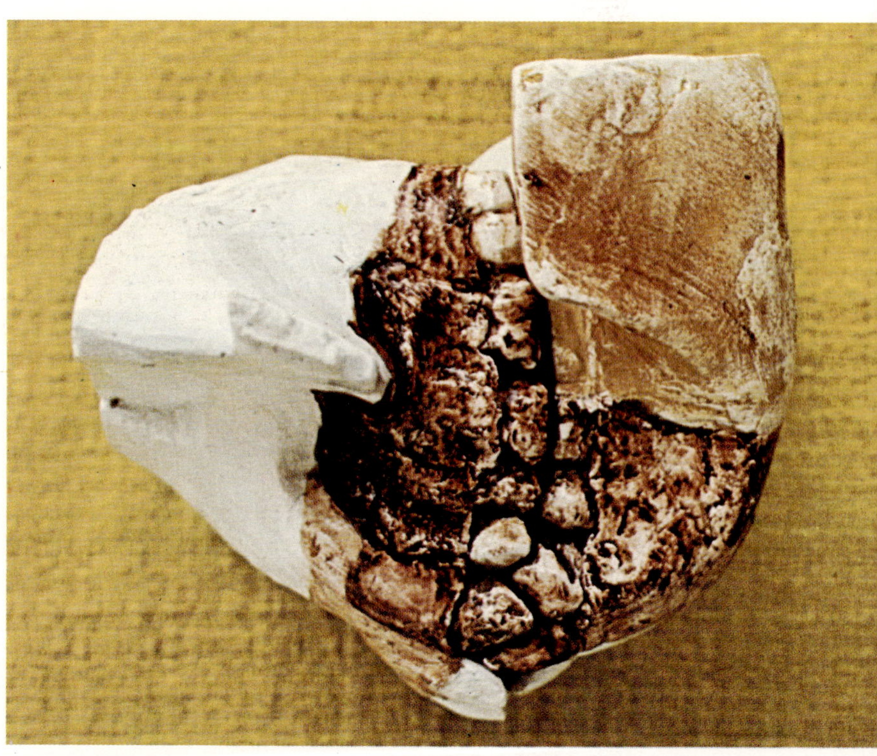

Right: side view of the jaw of *Ramapithecus*. The American paleontologist E. L. Simons classified both the Asian and East African primate finds as hominids. *Ramapithecus* has now been confirmed as the first demonstrably human type "from the subhuman chapter in the history of our descent, before the decisive transformation into man."

Above: Raymond Dart (right) with the much younger Richard Leakey. Dart, whose great find, the Taung child, came 35 years ago, was born in Australia and in 1922 went to South Africa as an anatomist. His claims for the significance of his discovery there two years later were dismissed at the time by the scientific community as those of an outsider bent on making a name for himself by producing a new and completely misleading theory.

Left: sifting the earth for fossils at a dig near the Olduvai Gorge. It was in this way that an African participant in Richard Leakey's expedition found fragments of a skull which turned out to be 2.5 million years old.

Chapter 7
The Incredible Dolphin

The ancient Greeks told amazing stories of communication between humans and dolphins, stories which were often dismissed as legend. But modern research suggests that they might have been true, based on what we now know of the high intelligence and affectionate temperament of this marine mammal. The resemblances between the brains of the two species, man and dolphin, are striking, but further studies of the dolphin's activities are plagued with difficulties: dolphins are elusive in their natural habitat, but on the other hand their behavior is likely to change if they are kept in captivity. However, scientists want to know much more, particularly about the dolphin's capacity for language—does its collection of mysterious noises indicate a complex pattern of speech?

In the 2nd century A.D. a dolphin was often sighted in the bay near the town of Hippo, a North African province of the Roman empire. At first everyone fled in fright, but eventually a young local boy made friends with it. Soon other children lost their fear and stroked the animal, which would swim confidently up to the shore, but the boy who had made the first contact with it was able to ride on its back. Later a second dolphin arrived which escorted the boy and his dolphin friend on their rides around the bay. The news of this strange friendship spread down the coast to other towns and villages, and many people, including distinguished visitors, came to Hippo to see the boy and dolphin playing in the sea. The people of the small town were at first delighted with their fame but eventually became disenchanted because the community could not afford to pay the expenses of the dignatories. Their quiet way of life was completely disrupted, so the city fathers had the dolphin killed.

This account, recorded by the 2nd century Roman writer Pliny the Younger, is just one of numerous tales of man's harmonious relationship with these apparently friendly marine creatures. Dolphins have been considered sacred by the Minoans of Crete and later by the Nabataeans. The Greek philosopher Aristotle observed a female dolphin giving birth and noted that, as in humans, the baby was born from a womb, suckled (underwater),

Opposite: Roman mosaic from Carthage, dating from the 3rd or 4th century A.D., depicting a nereid or sea nymph traveling on a dolphin. Greek and Roman mythology are both filled with stories of the intelligence and friendliness of dolphins in their relationships with humans and gods alike.

Above: bottlenose dolphins (porpoises) and their young in captivity at Marineland in St. Augustine, Florida. As with most mammals, the young dolphin stays with its mother for some time and is suckled on milk underwater. The mother's small nipples lie in a groove on the abdomen, and the baby feeds in short bursts as it must come to the surface to breathe every 30 seconds or so. For the first two weeks the calf is suckled about twice an hour, night and day, but by the time it is weaned, between the ages of 6 and 18 months, it is down to a mere six feeds per day.

received care and attention from its mother, breathed air through lungs, and uttered a variety of sounds. The Greeks revered dolphins, calling them *Delphinus* from *delphys*, meaning womb, and no Greek sailor would consider killing one unless faced with utter starvation. The Romans, however, considered dolphins just another source of food, and their name for the creature was porpoise, from the Latin *porcus* meaning pig and *piscis* meaning fish.

In Britain today the term porpoise is given to a beakless marine mammal, generally small in size (maximum 6 feet long) with a triangular dorsal fin and spade-shaped teeth. The seven species are members of the same family as dolphins, Delphinidae, and four are considered common porpoises—*Phocaena phocaena*, *P. sinus*, *P. spinipinnus*, and *P. dioptrica*. However, in the United States it is the bottlenose dolphin, *Tursiops truncatus*, which is called the common porpoise and which has figured in seaquarium shows, movies, and television programs. Most dolphins are characterized by a beaklike snout and slender, streamlined body, such as the common dolphin, *Delphinus delphis*. However, the family Delphinidae includes as many as 62 species, some of which do not fit the usual image of a dolphin: the killer whale, *Orcinus orca*, and pilot whales (*Globicephala*) are also members of the dolphin family.

Long ago Aristotle noted that dolphins make sounds, and in recent years "dolphinese" has been studied by many eminent researchers. Dolphins seem to "speak" to one another, and they send out signals that humans can hear. But since the members of the order Cetacea do not have vocal chords as do most mammals, how they make some of their noises is a mystery. Common por-

Porpoise–or Dolphin?

Left: a trained killer whale, *Orcinus orca*, performing at a seaquarium. Relatively little is known about their habits in the wild, but they are inquisitive and usually take a close interest in anything that might prove edible. The killer whale is a voracious feeder and will eat anything that swims. Included in its diet are whales, dolphins, seals, penguins, fish, and squid. Killer whales hunt in packs and will even take on huge blue whales, whereas they themselves probably have no real enemies aside from irate human whalers and fishermen.

How Do Dolphins Learn to Speak?

poises (bottlenose dolphins) were overheard first at a seaquarium in Florida and later in California and elsewhere. Eventually a wide vocabulary of audible sounds was noted. It had long been known that air could be released through its blowhole while a dolphin is submerged—one can see the bubbles rising like strings of crystal beads. In fact it is only after training that dolphins rise out of the water to "speak" in the open air. But concerning the squeaks, grunts, whistles, creaks, and "squarks" next to nothing is known.

There are theories that the sounds may originate either in the larynx region or the nasal passages, probably at two or three levels. Two tonguelike strips surrounded by powerful sphincter muscles form the epiglottis, and this may help in the formation of sounds. In 1975 research on the freshwater Indus River dolphin by G. Pillereri and his colleagues at the Brain Anatomy Institute of the University of Berne, Switzerland, revealed that its sonar transmissions ceased while it swallowed fish and began again shortly afterward. This finding supports the hypothesis that the transmitting organ is near the larynx. However, there are also valves controlled by large muscles in the nasal passages, and sounds may be produced here when the valves partially close the passages, causing air to vibrate audibly. It has also been intriguing to explore what the sounds mean. Do dolphins really speak to one another in their own language?

Although researchers have had to work solely with captive animals, reliable reports of observations made at sea do suggest that dolphins communicate by signals. Distress calls have been

Below: recording the sounds made by dolphins at Marine Studios. Underwater microphones pick up their squeaks, grunts, whistles, creaks, and "squarks" as well as the high-frequency ultrasonic sounds which are inaudible to the human ear.

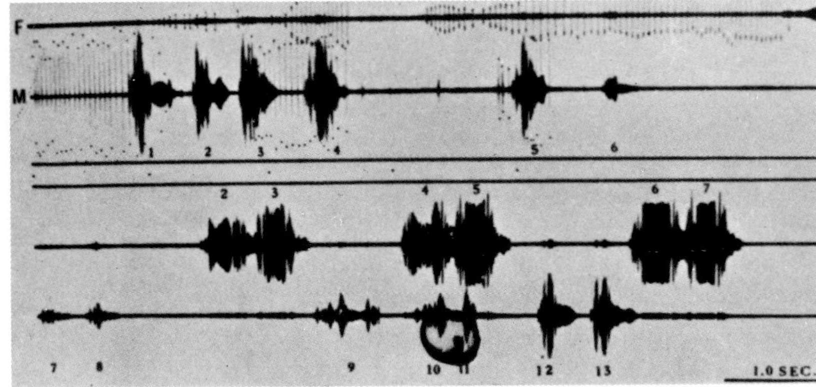

Left: a graph representing a conversation between two dolphins. The upper series of recordings shows an exchange of clicks and whistles, and the lower recordings show whistles without clicks. In each the upper line is the female's "talk" and the lower line the male's.

monitored from injured or endangered dolphins which bring other dolphins to the scene. Sometimes a single dolphin will leave a school, apparently to investigate an unknown object, reconnoiter a sea passage, or scout for danger, and it will then swim back to the group to report its findings.

One important discovery about the sounds dolphins make came to light during World War II, when the United States Navy instituted a method of forewarning its ships of the approach of enemy submarines. They used a *hydrophone*, an underwater microphone able to pick up sound waves over large distances in deep water. They named it SOFAR (Sound Fixing And Ranging) and were able to detect otherwise inaudible sounds over many miles. But quite frequently sounds were picked up that were not made by submarines or boats, and shortly after being picked up they would mysteriously disappear. It was suggested that the clicks and creaks might be due to animals such as dolphins or porpoises rather than the ship's propellers—surely the collection of sounds emitted by a passing school could make enough noise to confuse those listening to a hydrophone. In this way the sea was revealed to be a very noisy place, where prawns and shrimps cracked their claws like pistol shots and various fishes produced an array of sounds ranging from grunts and moans to clicks and chirps. Research since the war has linked the sounds with various aspects of animal behavior ranging from courtship to warning off enemies.

It was not until even later that the first hint came that dolphins might have their own navigation system. Arthur McBride of Marine Studios in St. Augustine, Florida, noticed that free-swimming bottlenose dolphins could avoid being caught in fine-mesh fishing nets (used to catch mullet and other fishes) by leaping over them. They would even do this at night or when the water was murky, so they could not have been "seeing" it visually. Could it be that dolphins had evolved a system similar to that of bats for finding their way around? Bats "see" in the dark by sending out very high frequency sound vibrations and then picking up any echo rebounding from whatever solid objects lie in their path.

This system is known as *echo-location*, and McBride suggested that dolphins might use it for navigation and finding food beneath the waves. Then in the 1950s several scientists working at different marine centers managed to record sounds on special tape recorders, which were capable of recording very high frequencies,

Below: a dolphin swims past an underwater microphone.

showing that dolphins were able to echo-locate using signals outside the hearing range of humans.

Dolphins produce sounds of over 100,000 cycles per second (100 kHz), high above the range picked up by the human ear (frequencies between 30 and 15,000 cycles per second). Water conducts sound waves well, absorbing far less energy from transmitted signals over a given distance than would be absorbed over a comparable route through air. Yet it is not necessarily a good medium for echo-location, because sound is conducted through water about five times more slowly than through air. However, the dolphin's system using very high frequencies seems to work almost perfectly. It produces sounds at a rate of 800 to 1200 clicks per second, faster than any muscle can produce them.

Much research has shown that dolphins and toothed whales such as killer whales do navigate using sounds and their hearing system. They can tell quite a lot about how far away and how large an object is. In tests dolphins have been carefully blindfolded using foam eye masks (so as not to irritate their sensitive skin) and yet they have been able to distinguish between 2.5-inch and 2.25-inch ball bearings from as far away as 10 feet. Further research has shown that they are capable of distinguishing between 6- and 12-inch-long fish. They can avoid wires as thin as 0.1 to 0.15 inch in diameter, and they can be trained to choose between targets varying in size by as little as 7 percent.

When a dolphin comes to the surface of the water to take a breath of air through the single blowhole on top of its head, the air circulates through its nasal passages under pressure to produce clicks and similar sounds. Behind the dolphin's domed forehead is a round mass of blubber called the *melon*, which is

Below: hosing down a gray whale (*Eschrichtius glaucus*) in a tank after it had become stranded on a beach. They are very slow swimmers and come very close inshore, which makes them vulnerable to hunters and leads to them becoming stranded at times. In the foreground are two dolphins.

Why Does Echo-Location Fail?

between the blowhole and the forehead. This seems to direct the sounds forward so that they reflect back from fish or other obstacles. The dolphin picks up the reflected sound waves in its lower jawbone, which transmits them to the inner ear.

A curious consequence of the system of echo-location is that schools periodically run aground. Individuals sometimes become stranded, but usually they are very old or injured. What causes these navigationally well-equipped animals to beach themselves? Even if people try to help by towing them back into deeper water, they invariably turn around and retrace their course.

Various suggestions have been made as to why these animals become confused and disoriented. One is that internal parasites eat their way into vital organs such as the brain or nervous system. The most plausible answer is that sound waves sent out by the dolphin or whale cannot be directed at objects below the line of the animal's jaw. Mishaps usually occur when the beach slopes very gently, and in these shallows the animal cannot determine slight changes in depth beneath it. Anyway, gently sloping beaches are not good at reflecting sound waves that do hit them, so it is possible that the cetaceans might panic at their disorientation and dart off as a group like a cattle stampede to end up on the beach. Another possibility, under circumstances where the sea floor is very muddy, is that the mud absorbs the majority of sound waves beamed at it so that only a very small proportion are reflected back.

Although some researchers have suggested that dolphins produce around 2000 different sound signals, most believe the number to be closer to 32. If each distinct whistling sound is a

Below: white-sided dolphin of the genus *Lagenorhynchus*, one of the 62 species of cetaceans in the family Delphinidae. Whales and dolphins have an insulating layer of blubber but no sweat glands, nor can they pant as dogs do, so other means are needed to lose excess body heat. The tail flukes and flippers are always warmer to the touch than the rest of the body, and their temperature is not only higher than that of the rest of the body but also varies over a greater range. These parts also have a much thinner layer of blubber, so it is assumed that whales and dolphins must keep cool by losing heat through their flukes and flippers.

Testing Dolphin Intelligence

complete expression, it would mean that dolphins are able to signal only 32 different situations. For example, one sound might mean "Help, here I am, come and find me," while another could signify "Hello, how do you do?". However, if each whistling sound stands merely for a symbol or word, then the animal might be able to combine them into words and sentences to form a complex language for communicating far more information.

The question remains as to what role language plays in dolphin behavior. Is it as complicated as human language, like the whistling languages used by some natives of Mexico, Turkey, and the Canary Islands? Or is their language comparable to the sounds and songs of birds or the barks and growls of wolves?

Many scientists believe that the dolphin's language is complicated because the brain of a dolphin compares closely in size and complexity with that of an adult human. It weighs more than a human's, on average 55–60 ounces (the heaviest recorded is 80 ounces), some 5–10 ounces more than a man's. Although dolphins are larger than men, the ratio of the dolphin's body size to that of its brain is only very slightly more than a human's. A dolphin's development from birth is very similar to man's, but the growth rate of the dolphin's brain is much faster than that of the human brain. Brain weight related to body weight or size can give scientists a fairly reasonable basis for comparing the mental capacities of animals. The table below shows average figures for man, the chimpanzee, and the dolphin.

	Man	*Dolphin*	*Chimp*
Length (in Feet)	5.7	8.0	2.5
Body Weight (in Pounds)	150	300	110
Brain Weight (in Pounds)	3.1	3.5	0.77
Brain as Percentage of Body Weight	2.1	1.17	0.7
Brain Tissue per Foot of Body Length (in Ounces)	8.5	7.0	2.5

These rough averages represent only one way of estimating comparative mental capacity, but some scientists do believe that the dolphin's brain is closer to man's in potential than is the chimpanzee's.

The surface area of the brain is thought by many to be a more relevant factor for comparison. Considering the cerebral cortex, which with the cerebrum forms the roof of the forebrain in mammals, that of dolphins—and indeed all cetaceans—is highly convoluted, with a complexity resembling that found in the brains of man and other primates. But although the dolphin brain is large, recent research has shown the neuron density to be relatively low. *Neurons* are nerve cells which are arranged in layers, each communicating with one or more of its neighbors by means of intertwined threads or processes (*dendrites*). The more neurons there are per cubic inch, the greater the intelligence of the animal.

Perhaps the major problem in solving the mystery of dolphin intelligence is that the concept is still so difficult to define and quantify. The "intelligence tests" of the early 1960s are no longer regarded as absolute measurements—they are now considered rough but fallible guides. Even the comparison of complex behavioral patterns is not valid in this case, since most detailed studies are performed with captive dolphins, and the artificial conditions no doubt influence their behavior. Perhaps the only certain point that can be made is that dolphins exhibit an intelligence which is at least equivalent to that of the lower primates such as monkeys and baboons, and perhaps even as high as apes such as chimpanzees.

Of the many tales of dolphins saving people from drowning one of the earliest and best known concerns the Greek poet and musician Arion, who is mentioned in Shakespeare's *Twelfth Night* (Act I, Scene 2). It dates from around 600 B.C., although the story was not written down until some 200 years later by the historian Herodotus. Arion was a native of the island of Lesbos, and he sang and played the harp more beautifully than any other musician of his time. He even traveled to Sicily and the Italian mainland where he amassed a wealth of gifts and prizes for his musicianship. For his return to Corinth, Greece, he boarded a boat at Taras (now Taranto) in southern Italy, but the crew turned out to be pirates. Once at sea they plotted to rob Arion and throw him overboard. The pirates knew that if Arion were allowed to reach Greece he would report them to the king, so they made him agree to throw himself into the Mediterranean after letting him sing one last song. His dying wish granted, Arion sang a high-pitched song addressed to one of the gods before jumping from the stern of the boat.

Once in the sea a dolphin came to Arion's aid and carried him on its back to Taenaros (now Cape Matapan), the southernmost point of mainland Greece. Upon reaching land Arion went straight to the king, who did not at first believe his story and kept him under guard until the crew arrived and confessed.

Below: watercolor by the 16th-century German painter and engraver Albrecht Dürer, illustrating the legend of Arion and the dolphin. The possibility of a personal and beneficial relationship between a human and a sea animal has intrigued people for centuries, partly as a fulfillment of the perennial human wish to be in close touch with nature.

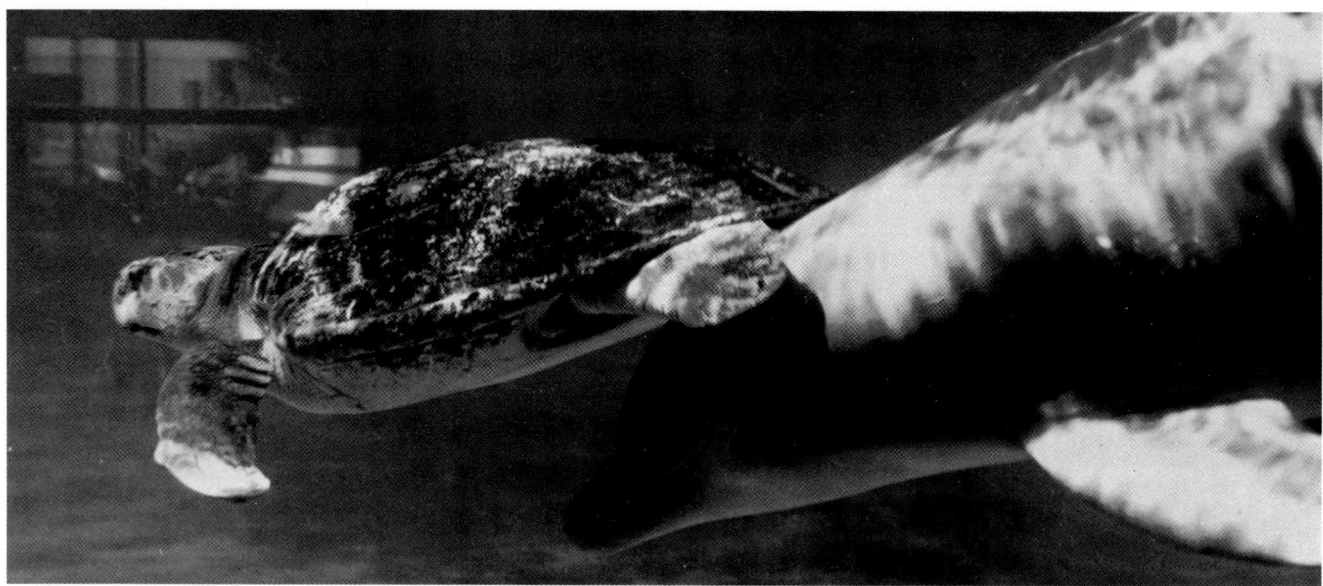

Above: a dolphin "supporting" a turtle at Marineland, in St. Augustine, Florida. This is probably the result of an instinctive behavior mechanism which encourages a dolphin to push up in order to keep a floating object on the surface.

Arion's story was retold generation after generation in poetry and song, and not long after Arion died the city of Taras minted coins decorated with the figure of a man riding a dolphin. This may have been because the city's legendary founder, Taras, a son of the ocean god Poseidon, had once been in danger of drowning at sea and had supposedly been saved by a dolphin sent to his rescue by his father. This could mean that Arion made up his own dolphin adventure after hearing the legend of Taras. In any case, it seems possible that at least one or two people have been rescued by friendly dolphins, since Greek literature abounds with stories of dolphin-human friendships. Later the Romans enjoyed telling stories which linked individual dolphins to admirable human qualities.

There have also been claims of dolphin rescues in quite recent times. One case occurred in 1943 on the coast of Florida and involved a woman who had gone alone to a private beach. When she was only waist-deep a strong undertow pulled her off balance, upon which she panicked, swallowed water, and began to lose consciousness. She remembers receiving a "tremendous shove" and found herself on the shore. On turning around she could see a dolphin leaping about in the water a few yards offshore. A passer-by claimed that he saw what had looked like a dead body in the water which the dolphin had pushed ashore.

What is the scientific evidence for this mysterious behavior of dolphins? Are they really intelligent enough to understand when a human being requires help or is it just the dolphin's natural playfulness and inquisitiveness? The most plausible explanation is that the act of rescue is the result of an instinctive behavior mechanism, which causes the animal to push in order to keep a floating object on the surface. It has often been observed that the first thing a mother dolphin does after giving birth is to push her baby to the surface and support it there so it can draw its first breath of air. Other dolphins often help in the task, and there have been several cases where dolphins have helped to keep afloat dead creatures or inanimate objects. One dolphin was seen pushing a waterlogged mattress ashore, while another kept the corpse of a

Amazing Rescues

tiger shark, its companion in captivity, floating on the surface of its tank for a week. The dolphin resisted any attempts by the keepers to remove the corpse until the shark began to decompose.

Dolphins do seem to use intelligence in their response to distress situations, thus bringing the instinctive behavior into operation. For example, when a strange male dolphin was introduced to a tank where a female dolphin was in distress and unable to rise to the surface to breathe properly, the stranger immediately responded by pushing the female up to the surface against the walls of the tank. Another female dolphin was brought in and the two healthy cetaceans did not leave the distressed animal for 48 hours. In the end they had worked out 10 different ways of supporting the sick female, the last ones involving a minimum of effort.

Over the last few decades there have been reports of situations similar to that of the dolphin at Hippo. In 1955 in Hokianga Harbor, northwest of Auckland, New Zealand, a man-loving dolphin appeared. It was soon named Opo because the creature first mingled with bathers near the beach at Omapere. Soon thousands of people were descending on Omapere, and the economy of the modest village boomed. One 13-year-old girl, Jill Baker, seemed to be Opo's favorite, since the female dolphin always sought Jill out from among the other bathers. Although a law was passed the following year forbidding the capture, hunting, or molesting of any dolphin entering Hokianga Harbor, Opo was inadvertently killed by fishermen using explosives to catch fish in the bay.

New Zealand's dolphin of the 1950s brought back memories of the famous Pelorus Jack, a Risso's dolphin (*Grampus griseus*) that for a quarter of a century accompanied ships through the Cook Strait between the North and South Islands of New Zealand. He did not make personal contact with human beings, but nonetheless he delighted the sailors on board the ships. In 1912 he mysteriously disappeared. He may have died of old age or from injuries received when the steamer *Penguin* accidentally rammed him. A man called Moeller found the body of a gray, 14-foot dolphin floating in the sea and brought it to officials of

Left: Pelorus Jack, a white Risso's dolphin which for nearly 25 years, between 1888 and 1912, piloted ships safely through the hazardous channel between New Zealand's South Island and D'Urville Island. In 1912 he mysteriously disappeared.

Adventures of Dolly, Nino, and Donald

the Wellington Museum. A ship's captain who had seen Pelorus Jack more than a hundred times proclaimed that this was not he. However, Moeller steadfastly maintained that he had found the carcass of the famous dolphin, and it is a fact that Pelorus Jack never reappeared.

In 1971 a bottlenose dolphin began to appear daily at the dock behind a house belonging to the Aylesbury family in the Florida Keys. They called it Dolly and she eventually became a pet, accepting fish from Mrs. Jean Aylesbury, escorting the two children when they went swimming, and allowing them rides on her back. After many inquiries the Aylesburys discovered that Dolly had belonged to the U.S. Navy and had been assigned to a dolphin-training base in the area. Regarded by her trainer as undisciplined and unreliable, Dolly had been released when the study center was moved. Dolly had by that time become conditioned to human companionship and communication, so she sought out new friends and found the Aylesburys.

In 1972 a dolphin was reported accompanying fishing boats and swimming among the bathers along the beach near La Coruña on the northwestern coast of Spain. First called Nino, the dolphin was later identified as a female and rechristened Nina. For five months she attracted tourists to the small town, and a statue was planned depicting a diver with his arms around a dolphin. But then the dolphin was found dead, and life for the humans returned to normal.

In 1972 a bottlenose dolphin also appeared at Port Erin on the southwestern tip of Britain's Isle of Man. For the next few years Donald appeared from time to time at various places, and he was sighted as far south as the resort of St. Ives in Cornwall, England. As in other man-dolphin associations he had his favorite people, and he showed particular curiosity concerning the activities of divers. One of his habits was lifting the anchors of small boats off the seabed, and on a few occasions he actually towed the boat away. In 1978 he was still a much-loved celebrity in the area.

The 20th century has seen for the first time humans slaughtering dolphins by the thousands. France once paid a bounty for every dolphin tail produced, and while permits for killing are no longer issued for French waters, other parts of the world still allow the practice. In some parts of the world whales are protected but dolphins are not, since they are supposedly in no danger of extinction. They are most frequently killed to provide meat for the pet food industry.

In the Pacific Ocean dolphins are frequently followed by tuna, which swim in slightly deeper water. They are not dependent on the same food source, so their association is a mystery. Modern fishing vessels catch over 45,000 tons of yellow-fin tuna each year, and dolphins are often trapped in the nets. It is estimated that about 250,000 die each year due to tuna fishermen, and between 5 and 7 million porpoises and dolphins have been killed over the last 20 years. The United States passed a Marine Mammal Protection Act which attempts to limit the maximum number of porpoises and dolphins killed yearly. In 1976 fishing operations were stopped when the figure of 78,000 was reached. However, fishermen from other countries such as Canada, Costa Rica, Mexico, Panama, and Nicaragua are still uncontrolled and can continue their activities. Scott Sinclair of the National Marine Fisheries Service states that "Eighty percent of the dolphins are killed by 8 percent of the boats." Another disturbing aspect is that most of the dolphins caught are females with their young. This has probably created a sexual imbalance in the dolphin population, since with more males than females their numbers may not increase for some time even if adequate protection is eventually enforced.

One of the latest massacres was in March 1978 when newspaper headlines around the world announced that over 1000 bottlenose dolphins had been clubbed to death by Japanese fishermen. The resulting worldwide outrage forced the Japanese government to invest almost half a million dollars over a three-year period in the search for a way to drive the dolphins from the fishing grounds. Included in their survey will be sophisticated deterrents such as high-frequency radio waves.

In the meantime the slaughter continues. The fishermen of Iki Island, Japan, have vowed to wipe out the dolphins—in April 1978 they stopped fishing for cuttlefish and yellowtail to chase a 1000-strong dolphin school. Many were trapped in shallow water and the throats of at least 20 were cut. The fishery cooperative on Iki Island announced that its members will continue to slaughter dolphins because they eat up to 40 pounds of fish a day and so threaten the fishermen's livelihoods. They have planned an "Anti-Dolphin Combat Year" for 1978.

Thus dolphins may soon be in as much danger as whales. The next few years will be crucial for the future of these species, while hopefully at the same time research into their language and intelligence may reveal answers to some interesting questions.

Below: Donald with a diver. After his first appearance off the coast of the Isle of Man in 1972 he began to entrance divers, boatmen, and fishermen with his friendly and playful habits. Insatiably curious, Donald frequently joined divers when they were performing tasks underwater. He even helped them or mischievously teased them by moving their equipment around. But usually when they stopped work and simply swam nearby he would become bored and swim off. Once he chipped a tooth while inspecting a movie camera when the motor was running. A television program was filmed about Donald in 1976 called "Ride a Wild Dolphin."

Chapter 8
Nature's Navigators

The average Sunday motorist makes his way across country with the aid of road maps, signs, and people he meets along the way. But all around him are animals making incomparably longer journeys across trackless oceans or dense forests without any navigational tools except their own instincts. How do they find their way, year after year? How do these animal migrators know when to set off, and when to return? How do they manage to feed and shelter themselves in unfamiliar territory? Do birds have a sixth sense that enables them to chart the earth's magnetic field? Can salmon taste the difference between different rivers or areas of the ocean? What part is played by the sun and stars? Scientists are still trying to solve the puzzles presented by nature's navigators.

If a man were blindfolded, taken away from his familiar surroundings, and released somewhere miles away from human habitation in an inhospitable desert or a steep mountainside, he might well die before he could find his way to other human beings. Among the few people who have survived this situation are the passengers of a plane which flew into a mountainside in South America. They survived by eating the corpses of those who had died in the crash, and eventually a team of two managed the descent and found help. Another was a young girl who became the sole survivor of a plane crash in the Amazon jungle. She had spent her early years in the tropics and knew something about surviving there, and after several days she stumbled upon a village. Parasites and grubs covered her open wounds and she was near death. She somehow managed to survive and was gradually restored to health.

Man is ill equipped to find his way around without maps, a compass, and other navigational aids. But all around him are animals which without apparatus or training are able to orient themselves and find their way home from wherever they are. The examples are numerous—cats and dogs have traveled hundreds of miles over unknown terrain to reach their master's new home; the arctic tern flies 12,000 miles a year from its winter feeding grounds in South America to summer breeding grounds

Opposite: hibernating ladybirds (or ladybugs), whose name dates from the Middle Ages when the beetles were associated with the Virgin Mary and called "beetles of Our Lady." The colorful species have a strong and unpleasant smell and taste equally bad. Their bright colors undoubtedly serve as a warning to predators. In summer ladybirds fly actively among foliage preying on aphids. In the winter they hibernate in large groups. They are very useful insects as almost all species prey on troublesome pests such as greenfly.

The Muttonbird's Amazing Journey

on the Arctic tundra. The navigational powers of animals are still mysterious to us, but over the last 50 years or so man has begun to fill in some of the blanks.

Migration is a round trip between two areas, both of which provide the animal with suitable conditions for survival at particular times of the year. The migration cycle which is established is regulated by the changing seasons. Some animals such as birds, insects, and bats travel through the air; antelopes, bison, and lemmings travel overland, and eels, salmon, turtles, whales, and seals travel through water. All of these animals are efficient migrators.

One migrating inhabitant of the air is the short-tailed shearwater, *Puffinus tenuirostris*, which breeds on the islands of the Bass Strait and along the southeastern coast of Australia. It is more commonly called a "muttonbird" in Australia and Tasmania, where huge colonies breed along the island's coasts. The name comes from the practice, dating from the beginning of the 19th century, of catching the young birds for food and as a source of oil and feathers. Some half a million birds were killed annually by 1910, but the population as a whole does not appear to have suffered from this heavy death toll. Today the muttonbird is the most common bird in Australia.

The amazing aspects of this bird's migration are the length of its journey and the route it takes. After spending the southern hemisphere summer, late September to April or May, forming pairs and raising a single offspring, they take off across the eastern Tasman Sea. Passing over the North Island of New Zealand, they then fly northwest toward the Japanese coasts. They reach Japan about a month after beginning their journey. They then continue on through the Bering Straits to settle finally on Wrangell Island, southeast of the Alaskan mainland. They remain until August or September, when they begin their return journey. They do not retrace their path but instead fly east and southeast along the Pacific coast of North America to California before striking a southwesterly course across the Pacific to northeastern Australia, after which they follow the coast south to their breeding grounds. Why should the species make this enormous figure-eight, covering some 20,000 miles, and repeat it with such regularity? The shearwaters arrive at their destination and begin to lay their eggs during the same 12 days at the end of November every year. Of the millions of eggs laid, the majority are deposited between the 24th and 26th of November. Everything is timed to happen correctly. After the breeding season is over the adult birds molt their body feathers in time for the northern journey, but they do not molt their wing plumage until settled in their winter quarters so that their wings are in the best possible condition for the second half of their trip.

The migratory loop seems to be adapted to fit in with the prevailing winds of the Pacific. No matter where they happen to be on their journey the prevailing winds blow in the direction of their flight, except over northeast Australia, where the shearwater must contend with the strong southeast trade winds that blow all year round. When casualties occur the victims are usually those birds making their first journey—those of only one year old.

Below: map showing the migrations of the short-tailed shearwater or muttonbird. The prevailing winds aid the birds in their circular tour of the Pacific after nesting in southeastern Australia. The recapture points shown are places where marked birds have been found in the process of studying the migratory route.

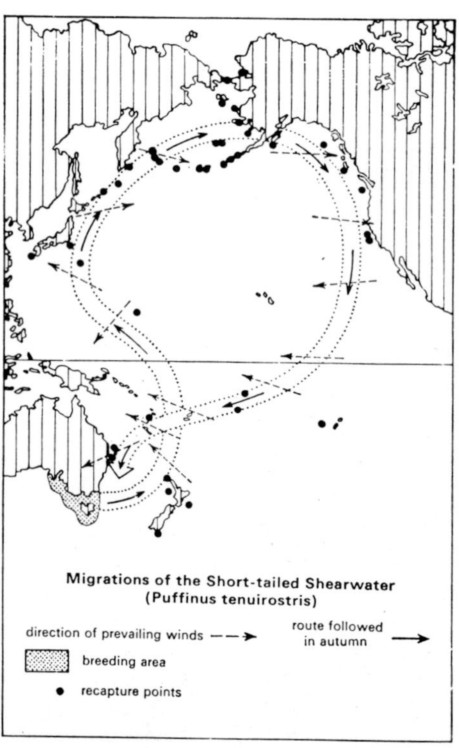

A more familiar example of fine navigation is the homing pigeon. Pigeon racing began as a sport around 1825 in Belgium, although the ancient Greeks, Romans, and Egyptians used pigeons to carry messages. Selective breeding over the last century has produced birds with amazing homing ability. But what many people do not know is that only 5 to 10 percent of racing pigeons make accurate, high-speed returns. Training them has become a fine art, one important aspect of which is to separate the male from its mate until it has found its way home. The homecoming rate is better if the female is incubating eggs and if the male knows what food can be obtained in its loft. Landmarks are very important. All birds have excellent eyesight, and during exploratory flights from the nest when young they learn about their surroundings. When released in unfamiliar territory they will fly around searching for recognizable landmarks.

But what about those birds that can return home from hundreds of miles away? This type of homing was demonstrated by an experiment with the Laysan albatross that breeds on the island of Midway in the center of the Pacific Ocean. Eleven birds were taken to different points on the fringe of the Pacific and released. All but two returned to Midway within 39 days, covering from 128 to 317 miles per day. Released in unfamiliar territory at different points of the compass, the birds made decisive moves to orient themselves and successfully returned to their nests. Research scientists have shown that a similar homing ability exists in other birds including the manx shearwater, black-headed gulls, swifts, terns, and starlings. They obviously possess a sense of

Below: the arrival of homing pigeons carrying vital dispatches during the siege of Paris in 1870–71. Pigeons have been used to carry messages since ancient times.

Biological Clocks

"locality" which helps them locate the place they were hatched, but many species also show a sense of "aerial navigation" where there are no land formations to go by.

Migratory birds have been attributed some rather fantastic abilities in the course of trying to explain their behavior. Some early theories supposed the birds could detect the intensity and direction of the lines of force which make up the earth's magnetic field. The bird would act as a conductor—as it flew over and through the earth's magnetic field, a small difference in magnetic potential would be created between the two ends of the bird's body. The animal would need to have the ability to interpret this information in order to pinpoint its position at any one time. This hypothesis has now been discarded. In experiments tiny magnets have been attached to birds which should, if the theory were true, upset the magnetic sensitivity of their bodies. However, the birds were still able to navigate just as well as before.

There must be some sort of compass within each bird that helps them get to and from their breeding and feeding grounds each year. In fact it is now known that every living animal has an internal rhythm which dictates the timing of its life. A daily life cycle lasts between 22 and 26 hours whether in an insect, crab, snail, or mammal, and it will repeat itself. This is called the *circadian cycle*, meaning "about a day." At first it was supposed that the periods of rest/activity were linked with the change from light to darkness, but intensive research covering all groups of animals has shown that the rhythm probably originates within the animal. This internal biological clock has been studied by scientists for many years, but as yet there is no clear understanding of how it works. It can be compared to a clock that keeps time through cogs, wheels, hairspring, and escapement. We understand a bit about the cogs and wheels but little of what triggers the hairspring. It must involve changes in the *permeability*, or penetrability, of an animal's cellular membranes, but that is as much as researchers know.

Below: map showing the migration of the indigo bunting (*Passerina cyanea*), which proceeds across a broad front. Buntings migrate from their wintering grounds in the Bahamas, southern Mexico, and Central America in early spring (late April) and arrive at their breeding grounds in the southeastern United States throughout the month of May. They depart south again in September and October.

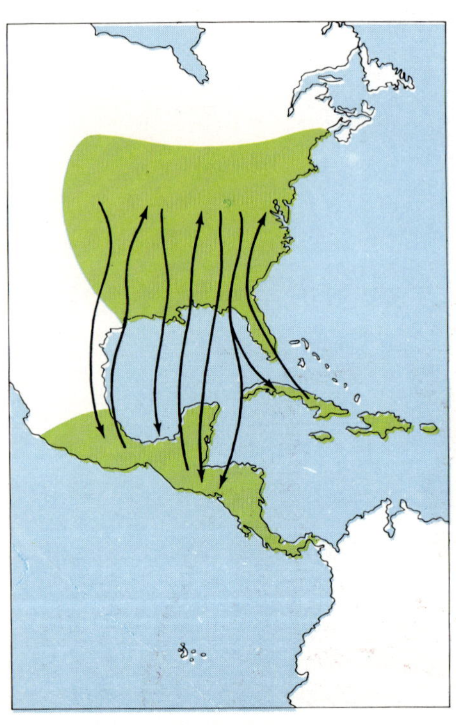

Recent research suggests that each individual living cell contains its own miniature clock. Evidence for this is the fact that when bits of tissue are taken from living insects or molluscs and are kept alive in a tissue culture, the individual cells still show a 24-hour cycle. Each animal contains millions of cellular clocks, all of them coordinated to produce a single living creature, organized and controlled in all aspects of its behavior. Further study has shown that there is a hierarchy of timekeepers. It is presumed that the brain is at the head of this hierarchy, partially because the brain is the only organ which receives immediate information from the eyes as to the external time of day. It can then send out messages to lower-placed clocks such as the adrenal glands which control many physiological functions, including the release of sex hormones, metabolism of carbohydrates, and the maintenance of the body's water balance. Adrenal tissue continues to release its hormones rhythmically when it is kept in a tissue culture. These lower-placed glandular clocks can inform the individual cellular clocks through the rhythmic release of hormones, which keep millions of separate clocks in phase with one another and with the external environment.

The concept of biological clocks is quite an adequate explanation for the rhythmic organization of normal body functions, but it is far too simple to cover all the various aspects of animal behavior. Most animal behavior patterns are inherited, although many are also inherently adaptable. Some types of behavior are constant—fleeing from enemies, eating when hungry, or coming into breeding condition. But even then the behavior pattern is not rigid. For example, if a bat is disturbed from its daytime sleep, it will fly away to another perch. Its biological clock would normally tell it to sleep and not to fly, but a more urgent stimulus overrides that instruction. When the danger is past the animal settles down to sleep again. Thus an automatic push-button control system would not be flexible enough, and its mysterious timing mechanism must be adaptable. One suggestion is that there is a kind of valve somewhere in the nervous system which can simultaneously control the level of response to different stimuli. Before it took effect, every response would be checked and given an all-clear. A strong enough stimulus would always get through. At the moment the construction of the valve is as mysterious as its workings, but it is probably situated in the brain, the processing powerhouse in higher animals.

The biological clock must play a part in the stimulation of migration behavior, and many researchers have experimented with artificial shifts in time. The indigo bunting, *Passerina cyanea*, migrates from the eastern United States to winter in Mexico, Central America, and the Caribbean. A research scientist called S. T. Emlen studied a flock in a planetarium in an attempt to discover how the birds navigate. Because they migrate at night he expected them to be affected by the position of the stars, but the birds ignored his artificial heavens. Outside it was spring, and the captive birds tried to fly north (which they would do in the wild), following their natural urge to return to their northern breeding grounds. Emlen then lengthened the "daily" periods of daylight to which the birds were exposed, but at a

Below: Stephen T. Emlen's experimental cage, in which he kept indigo buntings to test their system of orientation. Emlen studied a flock of birds in a planetarium and found that they adjusted their direction of flight not on the basis of star positions but depending on the amount of daylight. Their footprints leave a record of the birds' directional tendency.

Left: diagram showing the apparatus used in another planetarium experiment to investigate bird navigation. A replica of the night sky is projected on the domed roof while the human observer (symbolized by the eye) is enclosed within a tent of black felt cloth. Warblers were put into the cage when they were naturally ready to migrate. When the dome was filled with diffuse light only and no stars were visible, the birds were confused. When an image of the sky as it was outside was projected, the birds responded by taking up their usual migration direction. When the star pattern was altered so as to simulate different latitudes or longitudes, the birds altered their direction by the exact angle necessary to compensate for their apparent change in geographical location.

much faster rate than is normal for the coming of summer. In this way he altered the birds' internal clock to the point where they thought it was autumn, although outside their controlled environment it was still spring. The indigo buntings showed physiological changes associated with their autumnal southward migration. When exposed to a spring star pattern they were still determined to fly south as though it were fall. Their internal biological clocks, rather than the stars, were responsible for their changes in direction. Emlen then released his birds in a completely starless planetarium sky. They were completely confused and disoriented. When the star Polaris (visible in the northern hemisphere all year round) was replaced and the experiment repeated the birds regained their bearings to some extent. Thus it seems that birds are directed according to the seasons by their internal biological clocks, while at the same time they need a celestial point of reference, possibly Polaris, in order to navigate.

Birds that migrate by day appear to take a bearing on the sun to establish their route. This "sun compass sense" within the animal's body must be very finely tuned. As the earth rotates in its daily cycle the sun moves from east to west, and as the animal travels it has to be able to compensate for this movement. The birds must be equipped with an internal "compass" which, working with the sun compass sense, adjusts itself to the changing angle of the sun's rays. At every point on its journey the animal goes by local time.

The German scientist Gustav Kramer was the first, in the late 1940s and early 1950s, to study the problem of navigation by day. It had long been recognized that captive migratory birds became restless at the times of the year when they would normally be

Below: diagrams showing the experiments performed by the German scientist Gustav Kramer, using starlings. The birds were put into a circular container during daylight hours at the time of their usual spring migration. The container had six windows, each with a solid shutter bearing a mirror on its inner surface. The bottom of the container was transparent so that Kramer could record the position of each bird. When all six shutters were fully open (top) the starlings took up a position facing northwest as they would normally. When the apparent angle of the sun's rays was altered by the use of the mirrors (two lower diagrams), the starlings altered their position according to the angle of the light reaching them. If the light was made very dim and even throughout the container, the starlings were unable to orient themselves.

How Do Migrators Find Their Way?

migrating, although in other respects the birds were well adjusted to their restricted way of life. Kramer noticed that in spring the starlings he was observing became restless and spent most of their time in the northwest corner of their cages. He placed a screen around the cages so that the birds could only see the sky. Their behavior did not change, nor was it affected by the presence of large electromagnets. Here was evidence suggesting that the earth's magnetic field did not influence them. It seemed that the birds were using the sun to fix an angle of orientation. Kramer then constructed a special cage with mirrors. When placed in it the birds could only see the sun's reflection in the mirrors. They altered their bearings to coincide with this new position, and then readjusted them when the angle of the mirror was changed. The ability to orient themselves, whether by the stars at night or the sun during the day, thus seems to be inborn in migratory birds. The biological clock is effective even in young birds.

But birds are not the only navigators of the air. Many insects, moths and butterflies in particular, carry out annual migrations. Even those insects which do not migrate do leave the place where they spent their formative stages. Although these may not require the ability to navigate, certainly all flying insects need a way to orient themselves in order to hold a steady, undeviating direction of flight.

One important factor in the study of migration is that the eye of an insect is not the same as that of a bird. The eyes of an insect are called *compound eyes*, as each is composed of many light-sensitive cells called *ommatidia*. The exact number depends on the species; the queen ant's eye has only 200 facets, while in a dragonfly's there are up to 28,000. Between these two extremes

Left: the compound eyes of the horsefly (*Tabanus autumnalis*), anthropod eyes subdivided into many individual, light-receptive elements each including a lens, a transmitting apparatus, and retinal cells.

The Dancing Honeybees

there are insects such as the housefly with about 4000 facets and the butterflies, which have between 2000 and 17,000 according to the species.

Although we know the structure of compound eyes in insects, nobody really knows what sort of sight the insect obtains from each ommatidium. It is generally thought that the many miniature lenses build up a picture similar to a mosaic in the brain of the insect. However, each lens has a visual range of 20 to 30 degrees, so there must be considerable overlap between neighboring lenses. Does this not confuse the insect? The fact is that although the brain of most insects is only the size of a pinhead, it is sophisticated enough to inhibit the signals from adjacent ommatidia in such a way as to prevent them from becoming superimposed and confused. A network of nerve fibers connects the brain to the compound eye.

One of the most interesting aspects of orientation and navigation in the insect world is the food-finding technique of the honeybee. The well-known Austrian zoologist Karl von Frisch of the University of Munich spent many years, from around 1915 to the late 1940s, studying bees. His aim was to discover how bees communicate in their hives. In experiments Frisch put sugar water some distance from the hive while through a red pane of glass he observed the inside of the hive and noted the behavior of marked bees returning from the food source. They performed mysterious dances watched by many other worker bees. If a bee had returned from a food supply 10 yards from the hive it did a circular dance, reversing direction regularly at the same point in the circle. When the food was at the much greater distance of 1000 yards a figure-

Below: honeybee (*Apis mellifera*) collecting nectar. A worker bee is ready to go out foraging for nectar, pollen, water, and resin at the age of three weeks. She is guided by her senses of smell and sight. Foraging is very hard work, and after two to three weeks of it the worker dies. Workers hatched in the autumn, however, have a much longer lifespan, as they build up food reserves in their bodies and huddle together in a mass to keep warm.

eight dance was performed. The worker bee looped to the left around one of the circles forming the figure-eight and to the right around the other, with a "tail-wagging" run along the intersection of the two circles. "Near" food supplies were indicated by the circle dance, while "far" distances were indicated by the figure-eight dance. But how did they communicate the exact distance to be traveled? Frisch noticed that the "far" signal dance was performed at varying speeds. For a food supply at a distance of 100 yards the bee outlined five complete figure-eights in 15 seconds, while for a food supply 2 miles away only a single figure was danced in the same amount of time. This was fine, but how did the worker bees communicate the information as to which way to fly? For "far" food supplies the "tail-wagging" part of the dance varied with each of two different factors: the direction from the hive to the food and the time of day. For example, tail-wagging while ascending vertically along the honeycomb meant that food was available to the east in early morning, to the south at noon, or to the west as sunset was approaching. In every case the angle between the tail-wagging run and the vertical line of the honeycomb corresponded to the angle the bees should fly from the hive, using the sun as a point of reference. In the "near" circle dance the point at which the insect reversed its dance and the vertical line of the honeycomb gave the angle corresponding to that formed by the food, the sun, and the hive.

But how do bees actually control the direction of their flight? The bee's brain seems to be able to measure the angle at which the sun's rays strike its compound eyes, and it is thereby able to determine in which direction it is flying. If the sun is obscured by clouds but there is a patch of blue sky the bee can still find its food. How could this be if it used the sun as a reference point for navigation? Frisch was fortunate in that other scientists had previously discovered polarization. He was thus well aware that the angle of polarization of the scattered light from a seemingly blue sky varies greatly. Man can see this only through Polaroid material or a Nicol prism. But was it possible, asked Frisch, that the compound eyes of insects could detect this variation? In ingeniously simple tests with Polaroid sheeting he showed that bees were indeed taking cues from the polarized light of the sky as though it were a compass, so that it did not matter if the sky was overcast. The bee used its internal clock to compensate for the movement of the sun from east to west during the day. Since then other researchers have shown that many insects use this navigational technique.

Other amazing facts have been discovered about the bee in flight. An object first seen by one facet or a group of facets of its compound eye will be picked up by another a split second later. The bee's minute brain is able to measure this difference and thus to calculate its ground speed when flying. It can also measure its air speed, by noting the angle at which its feelers are bent by the wind as it flies. With this degree of sophistication it is no wonder that the bee can calculate an exact course between a feeding source and the hive.

With the aid of these abilities many insects navigate over not just a few miles but hundreds and even thousands of miles during migration. A fascinating spectacle was observed in the early

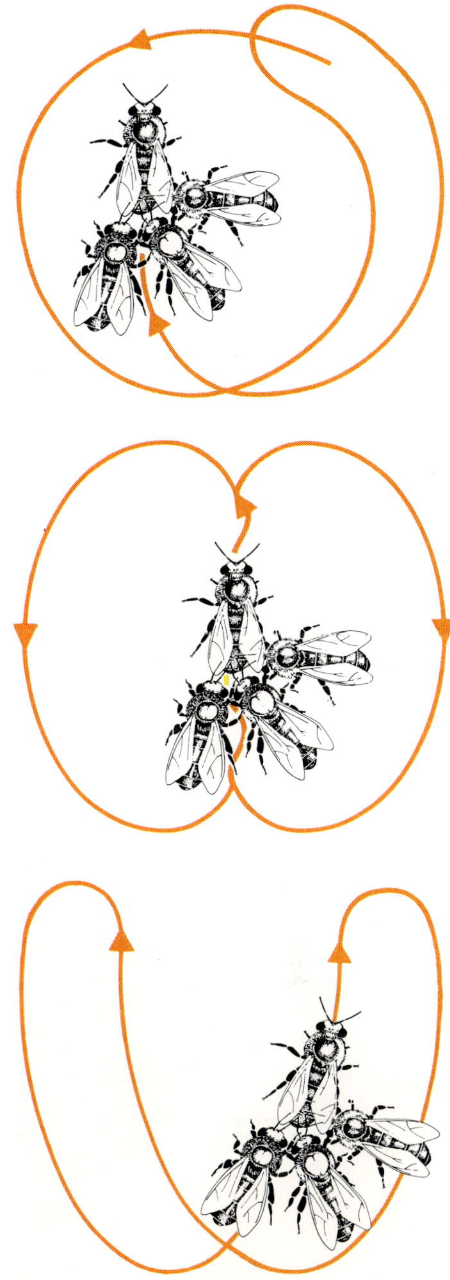

Above: how bees communicate the whereabouts of a food supply by dancing. The black Austrian honeybee (*Apis mellifera carnica*) describes where to find food less than 275 feet from the hive by performing a round or circular dance (top). If the food is further away, she performs a "waggle" dance or figure-eight (middle). The number of times the pattern is repeated within a given length of time indicates the exact distance of the food source. The Italian honeybee (*Apis mellifera ligustica*) does a sickle dance (bottom) for sources beyond 30 feet. The opening of the "sickle" faces the food source, and the intensity of the dance indicates the quality of the supply.

Trees Full of Butterflies

Above: butterflies (*Belenois aurora*) roosting in high grass for the night during their annual migration, when they cross southern Africa.

1920s by the British entomologist C. B. Williams. He found that three different populations of migrating insects flew over the Masai plain in Tanzania, East Africa, at the same time. A species of small yellow butterfly, *Terias senegalensis*, was flying southwest and large yellow butterflies, *Catopsilia florella*, were moving north-northeast, while millions of locusts, *Schistocerca gregaria*, were heading southeast. The three streams of traffic mingled about 10 feet above the ground, but even in the visual confusion not one collision was observed, and none of the streams were thrown off course by the rival migrators.

The best known migrating butterfly is the monarch or milkwood butterfly of North America, *Danaus plexippus*. It not only migrates seasonally from southern Canada to winter in southern California, the southern states, and Mexico, but the species has also spread westward to establish itself in Hawaii, Tonga, Samoa, Tahiti, New Zealand, Australia, and the Malay archipelago. This dispersal is due to a combination of long-distance shipping and meteorological conditions, when the migrating butterflies are blown off course by strong winds and eventually become established locally. They can only settle where the milkweed plant grows, as it plays host to the monarch's eggs and also nourishes the caterpillars. Entomologists have shown that those monarchs which reach Mexico breed there, but those wintering further north in California, Florida, and Louisiana merely hibernate in the trees. The same trees are used year after year by huge swarms, and in places such as Pacific Grove, California, thousands of tourists come to see the butterfly trees.

By marking and recapturing individuals entomologists have

Left: best known migrating butterfly, the monarch (*Danaus plexippus*). Mating monarchs gather in profusion before making off in couples. They also settle on trees in great numbers each night during their migration. They spend the winter in a state of semihibernation clustered on trees in Florida and southern California. The same trees are used as resting places each year.

Left: diagram showing the life cycle of the painted lady or thistle butterfly (*Vanessa cardui*). It is the only butterfly with a worldwide distribution (excluding only South America) without the formation of any well-defined races or subspecies. The explanation is in its urge to migrate, which is so powerful and persistent that its populations are subject to constant mixing, thus preventing the formation of local races. Large swarms of these butterflies travel regularly in the spring from North Africa to Europe and from Mexico to Canada. A return migration has been observed in autumn over Europe, but there is no clear evidence for a return journey across the American continents.

The eggs (1) are laid singly on thistles or nettles and hatch eight to nine days later. The caterpillar (2) feeds on these plants within silken webs which it spins among the leaves. After several weeks, the caterpillars transform into the chrysalids or pupae (3). These hang from the leaves within tents made from a single leaf spun together with silk. In about two weeks the adult butterflies emerge and, after drying their wings, fly off (4). The dates shown relate to those painted ladies that breed in Europe as they migrate northward from Africa, reaching northern Europe in early summer. Many of the adult butterflies are on their way back to Africa by October. The rest soon die, unable to withstand the cold, damp northern winter.

shown that some monarchs make a complete round trip—a true migration. They are accompanied by many others, the result of breeding en route.

Several other butterflies and moths make the trip south in autumn and north in spring, including the red admiral, the small tortoiseshell, and the large cabbage white of Europe. However, since most insect life cycles come to an end in less than a year and the adult insect usually dies after breeding, the offspring are often left behind in mid-migration to complete the journey and perpetuate the species. The new generation may return to the area its parents came from or remain in their birth area and even breed there. We do not know why insects migrate, but their internal biological clock is certainly involved.

Right: the death's-head hawk moth (*Acherontia atropos*), so named because of the startling skull-like marking on the back of its thorax. It is a heavy but fast-flying moth that occasionally travels from North Africa to northern Europe, sometimes in large swarms. This may be due to their being blown toward France and Britain by strong winds.

Ladybirds vs. Greenfly

In 1910–12 American entomologists wanted to turn the liking of the ladybug (or ladybird) for aphids, especially greenfly, to the advantage of California fruit growers. They found that under normal conditions the ladybird, a member of the Coccinellidae family, migrates to the mountains in the north of the state in autumn where it hibernates in groups of up to several thousand under leaves or stones or on the bark of trees. When spring arrives the insects fly 100 miles south to the fruit orchards, where greenfly have in the meantime been multiplying at a terrific rate and causing great damage to the potential fruit crop. To try to protect the fruit millions of hibernating ladybirds were collected—one man can gather a million in a day—and put in cold storage until the greenfly had got out of hand. Some 49 million ladybirds were then set free in highly infested orchards, and because one ladybird can eat about 60 greenfly a day the experimenters thought the greenfly problem would soon be solved. However, this seemingly ideal biological arrangement was an utter failure. Within two days the millions of ladybirds had vanished from the orchards, leaving the greenfly in control. This baffled the scientists until seven years later when similar experiments were performed, and the answer was only discovered then because the entomologists had the foresight to mark their experimental ladybirds. This time the vanishing population was found 100 miles away from the orchard, in a place approximately the same direction from the orchard as the orchard was from the mountains. What had happened? The answer was that the scientists had not taken into account the insects' biological clock and orientation system. Awakened from hibernation away from their normal mountain habitat, the insects still tried to make their return spring trip from the mountains where they thought they were to their summer feeding and breeding grounds. The innate rhythm of their migratory cycle told them which way to fly and this is what they did, arriving 100 miles away from the orchard in what proved to be an unsuitable habitat. The majority perished from a lack of food.

It is perhaps more difficult to explain the navigational techniques of the last group of airborne migrators—mammalian bats. Moving at night, several bat species journey hundreds of miles to find a suitable climate, hibernating quarters, or more food. In North America, for example, the red bat, *Lasiurus borealis*, migrates from the northern United States and Canada to the southernmost parts of the U.S. mainland, some even flying far out over the Atlantic Ocean. Few have been observed making the return trip, but the fact that they are again present in their northern habitat for the summer proves that they must have made it.

In Europe the majority of bats hibernate through the winter, but some travel across the Mediterranean to winter in North Africa. In the tropics many of the large, fruit-eating bats migrate according to the ripening times of different fruits. In Australia many thousands of gray-headed bats hang during the day in colonies, called camps, in the forests. Huge flocks move from camp to camp until they reach the extreme southeastern tip of Australia where they spend the summer.

It is well known that bats use not their eyes but high-pitched

Below: Mexican freetail bats of the genus *Tadarida* clinging to the roof of a cave. Bats are the only true aerial mammals. Several of the world's 800 or so different species fly great distances of up to 1000 miles each way to find a better climate, more food, or suitable hibernating quarters. It is quite difficult to observe bat migration because it only occurs at twilight, and in most cases only small numbers are involved at any one time.

Using Radar to Track Moths

squeaks and their echoes to locate obstacles and to pinpoint flying prey such as moths. Researchers have shown that many species are able to return to their roosts within a few hours from distances of up to 20 miles and within a few days from 200 miles away. But do they use their eyes to locate landmarks? Experiments in Trinidad in 1966 revealed that spearnosed bats could find their way home blindfolded from 8 miles away by using sonar. At greater distances of up to 20 miles, however, they did need to orient themselves visually. Migrating bats probably use a combination of these methods to navigate.

When it comes to other night-fliers such as moths little scientific work has been done because they are very difficult to track.

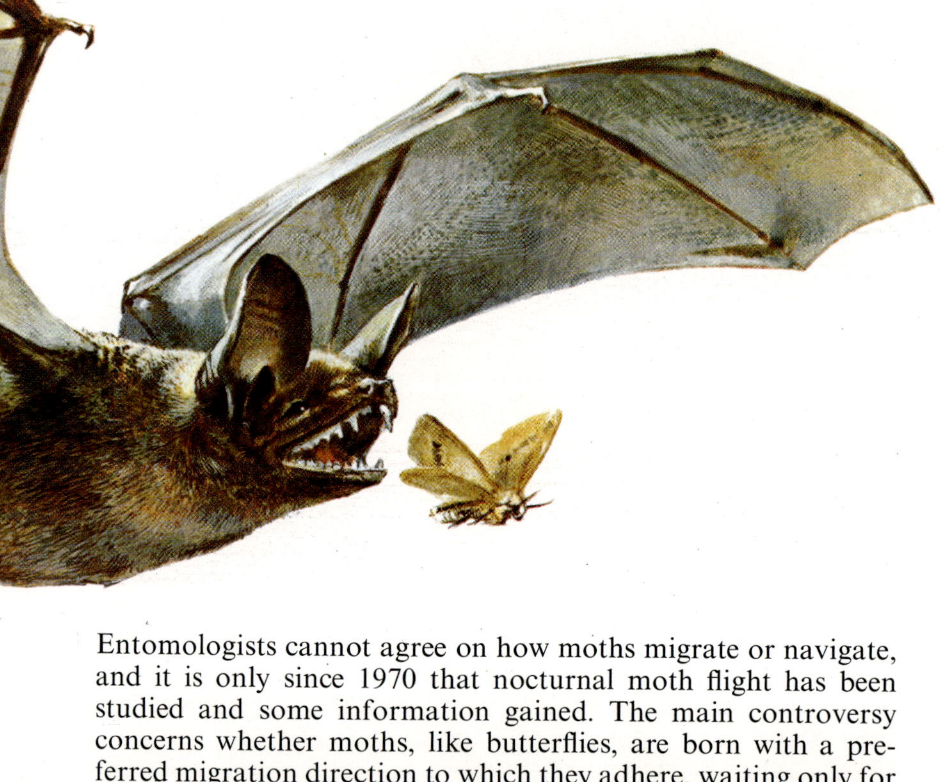

Above: a mouse-eared bat (*Myotis myotis*) about to catch a moth. It is among the largest European bats and is one of the strongest fliers. It makes a regular annual migration from its summer breeding grounds throughout Western Europe to the caves of the Atlas Mountains in Morocco. There the bats spend the winter in hibernation, hanging by their curved claws in huge colonies of up to 100,000 bats. When spring arrives the bats wing their way by night back to Europe. Bats show an amazing ability to remember features of familiar surroundings, but much less is known about how they find their way over long distances.

Entomologists cannot agree on how moths migrate or navigate, and it is only since 1970 that nocturnal moth flight has been studied and some information gained. The main controversy concerns whether moths, like butterflies, are born with a preferred migration direction to which they adhere, waiting only for the right tailwind, or whether they simply take off downwind irrespective of direction. Radar has been used to track these nocturnal fliers through arid parts of tropical and subtropical Africa and Australia. It appears that the majority take off one hour after sunset and rise rapidly to a height of about 1300 feet but definitely no higher than 5000 feet, above which the temperature is too low. Over the next one to four hours there is a gradual decline in height of the highest fliers. How long each individual moth spends in the air or how far each one flies have not yet been determined. Depending on conditions the insects stay in the air for between one and twelve hours and fly at a ground speed of between 5 and 50 miles per hour. The larger moths appear to travel downwind for most of their flight.

The findings suggest that the migration of moths, though it takes place only at night, is no different to that of butterflies. But what orientation cue do they use if there is no sun or polarized light to guide them? Until as recently as 1977 this was a complete mystery—the breakthrough was an experiment in Britain using

flying moths of the species *Noctua pronuba*. When kept in a totally dark room or when the moths' eyes were completely painted over they seemed to have no recognition of compass directions. This showed that neither wind nor the earth's magnetic field were involved. Then only the upper two-thirds of the moths' eyes were painted over, but they still could not orient themselves. However, when only the lower two-thirds of the eyes were painted the moths could get a bearing. Thus it was discovered that the moth does use sight. When the night sky is completely overcast the moth cannot orient itself, but when it is merely the moon that is not visible the insect appears to orient itself by using a star or star pattern. The findings show that a moth is normally able to adjust itself by 16 degrees every hour, even at night.

One of the most awe-inspiring sights of the biological world is that of a moving locust swarm. To be totally immersed in an unending cloud of flying, hopping, feeding insects is quite frightening. Locusts are short-horned tropical or subtropical grasshoppers, members of the family Acrididae. Not all locusts migrate—only a handful of species have evolved migratory behavior—and the various trigger mechanisms are by no means completely understood, as shown by the fact that locust migrations still cause crop devastation and famine in Africa, Asia, and even Australia.

Above: photograph of a swarm of locusts passing over Timbuktu, August 1895. The greatest depredations are primarily due to two species, the desert locust (*Schistocerca gregaria*) and the migratory locust (*Locusta migratoria*). In the Book of Exodus is a graphic description, written around 1500 B.C., of the arrival of a plague of locusts in Egypt, helped on their way by a strong east wind. When the great clouds finally left the land, "there remained not any green thing in the trees, or in the herbs of the field, through all the land of Egypt."

Plagues of Locusts!

Whether a locust species migrates or not is linked to the amount of available food and living space in its locality. If there is abundant vegetation and living conditions are relatively uncrowded the locusts look and behave like conventional short-horned grasshoppers. If, however, they hatch into an area of sparse vegetation where innumerable locusts are crowded together, the young grasshoppers hop for long distances in search of food. These "hoppers" join together in bands and move across the land devouring any vegetation in their path. Within the space of some 15 miles they will have gone through their final molting stage and their wings will be revealed. By this time their coloring has changed markedly from that of the non-migratory members of their species. The moving locusts have become bright—pink, yellow, and red—whereas the non-migratory specimens remain well-camouflaged by subdued greens and yellows. The migratory insects are active mainly by day rather than by night, have much longer wings than the others, and are able to delay reproduction for up to six months. For years it had been assumed that the two forms were distinct species, but modern research has shown that there are two different sets of characteristics, the existence of which depends on the conditions under which any particular insect group develops.

The most familiar of the nine main species of migrating locust is the desert locust, *Schistocerca gregaria*, found in the desert belt of North Africa, Arabia, Iraq, Persia, Pakistan, and northern India. Others include the South American locust, the migratory

Opposite: brown locust hoppers (*Locustana pardalina*) of southern Africa.

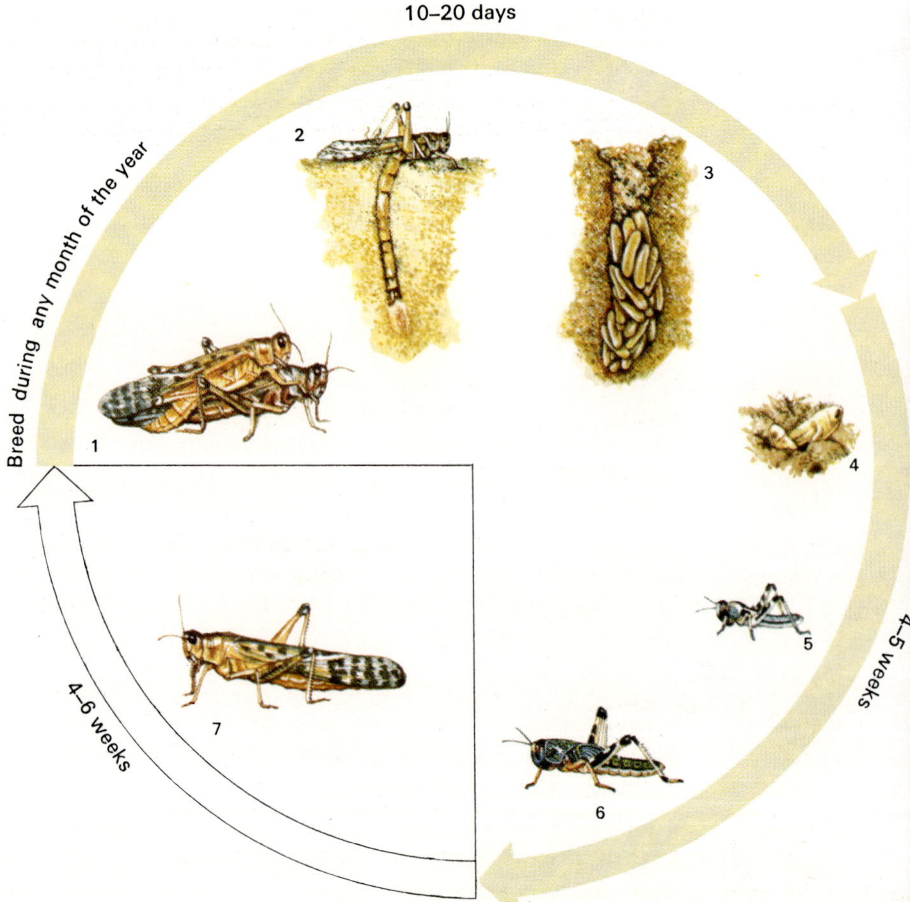

Right: diagram showing the stages in the life cycle of the gregarious phase of the desert locust. Desert locusts can breed during any month of the year. They turn from the pink color of immaturity to a yellowish color before gathering together in large, closely packed groups to mate at the start of the rainy season (1). The female locust pushes the hind part of her body into the desert soil until she reaches a moist layer, where she lays her eggs (2). She also produces a frothy secretion which hardens to form an egg pod containing between 80 and 100 eggs. Each female lays two or three of these egg pods, and each egg is the size of a grain of rice. The hole containing the pods is sealed over with the quick-drying froth and the eggs are left to incubate (3). The egg stage lasts from 10 days to 70 days, depending on the temperature outside. On hatching, the tiny young (4) tunnel to the surface and shed a membrane enclosing them. The young of solitary-phase locusts turn green within a few hours and disperse, remaining as solitary grasshoppers unless certain conditions, such as weather and food supply, force them to crowd together, when they change into the gregarious phase. The young of the gregarious-phase parents shown here turn black soon after hatching. After molting they are adorned with black and yellow stripes (5). These second-stage hoppers have no wings, but advance over the ground in huge bands. After passing through several further molts, during which the bands of hoppers march across the country, eating thousands of tons of green plant food on the way, the fifth stage is reached (6). Now the hoppers start to develop wings. Eight to eleven weeks after hatching they develop into immature pink adults (7). Aided by their newly found powers of flight, they travel in huge swarms to areas where they will breed, making journeys of up to 2000 miles or even farther. It is during this migratory period that the locusts do the most damage to crops.

Bison, Frogs, and Caribou

locust, the Moroccan locust, the Australian plague locust, the brown locust, and the red locust of the southern half of Africa. How locusts maintain their preferred compass direction and how they orient themselves are still unknown. The fact that locusts migrate in the same direction whether it is morning or afternoon implies that they must be able to compensate for the sun's movement from east to west across the sky, but that is almost all the information we have. It does seem, however, that whether butterfly, locust, moth, dragonfly, or ladybird, whether nocturnal or daytime flier, the orientation methods and patterns of migration are very similar.

Those mammals which migrate on foot are found primarily on the huge continental land masses such as North America and Africa. For several reasons North America has the greatest number of migratory mammals. The climate there varies greatly, and the main mountain masses, such as the Rocky Mountains and the Appalachians, run in a north-south direction and therefore would not necessarily be barriers for traveling animals. Regarding the other major land masses, Asia's mountain ranges run east-west and Africa has no extremes of cold, the climate ranging only from subtropical to tropical.

One of the most impressive migrations in North America was once the annual movement of the bison. Millions of bison migrated south from their northern summer pastures when cooler weather came. However, this spectacle is now a thing of the past. After the west was opened up in the 19th century the bison's prairie habitat was lost to ranching and farming, while whole herds were killed off as the railroad system was extended. It appears that the bison did not simply move to and fro in a north-south direction but in a more or less clockwise path—different herds used different paths. Most herds moved a total of 200 to 400 miles during the fall to reach their southern quarters.

Right: bison-hunting on the Kansas Pacific Railroad in the mid-19th century. Many of the westward-pushing railroads followed the well-worn bison trails, and special trains full of hunters were operated to provide track-laying gangs with fresh meat.

Although the bison are now gone, the caribou, North American counterpart of the Eurasian reindeer, still traverses a counterclockwise course during the winter before bearing off northwest in spring to the barren lands of the Arctic Circle. Their migration is partially connected with available food, but they may also be escaping the ferocious mosquitoes that can make their lives a misery. It seems that about 14 days before the mosquitoes hatch the caribou somehow receive a message telling them to begin their northern march. They thereby avoid the worst of the gigantic hordes of mosquitoes which ravage the more southern mammals. The caribou breed in the Arctic and usually return south after the last mosquitoes have died in the colder weather of the autumn.

In the 1800s the springbok, a small graceful gazelle, inhabited southern Africa in herds of up to 8000. Periodically, they migrated from the Kalahari desert in southwest Africa to the higher ground of the Cape in search of water as their habitat became drier and their numbers overstretched the food supply. Eventually, however, settlers cultivated much of the land and fenced it off, and springbok were shot for food and for sport. The last recorded springbok migration was in 1896, and today they are confined, in much smaller numbers, to southwest Africa.

Minor migrations still occur in other parts of Africa. The majority of grazing mammals, including the wildebeest, gnu, and common zebra, follow the rains for the new grass. These movements are irregular, though, and although extensive are ill-defined. Many zoologists consider this type of annual movement a seasonal shift of habitat rather than a true migration.

Many other land animals migrate with the seasons, but on a much smaller scale. Red deer of Europe and Asia move down from the hills at the onset of winter to sheltered valleys and the better food supplies there. It may be that migratory land mammals possess some sense of direction that we humans lack—how else could cats and dogs return to their homes after being removed in closed containers, since memory of the route would not be a factor?

Frogs and other amphibians find their way each year by some unknown method to pools and ponds where they congregate in large numbers and reproduce, only to disperse afterward. In North America many snake species travel each winter to the same "dens" to hibernate year after year. They probably navigate using their extremely acute senses of smell and taste—the forked tongue picks up small particles and a special organ situated in the roof of the mouth interprets their scent and taste.

Apart from air travelers the major migrations occur in the aquatic environment. Fish and sea turtles perform more seasonal migrations than any other animal group. This is probably because their watery habitat is constantly moving and if the animals did not move to more favorable feeding and breeding grounds they would be moved from place to place by the force of the currents. Migration and navigation techniques in the fish world can be complex because some species may move from fresh to salt or brackish water. Fish which breed in fresh water and then travel downstream to mature at sea before returning to the river to spawn are said to be *anadromous*—the best known examples

Below: frogs (*Rana esulenta*) spawning. Entire populations, usually spread over a relatively wide area, congregate in special places to breed. They are guided in their migration to their accustomed spawning areas by an extremely well-developed sense of direction.

Above: green sea turtle (*Chelonia mydas*), which has been wiped out over many parts of its range and made very rare in others because its flesh and eggs are good to eat and easily gathered. Its name comes from its green-tinged fat. Along with the other species of sea turtle, the green turtle migrates to its breeding grounds regularly. Preliminary experiments have been made in which radio transmitters are attached to certain female green turtles. As they swim slowly and are often not far below the surface of the water, they may even be able to tow a small float containing a larger transmitter tracked by satellite.

are the Pacific and Atlantic salmon of the family Salmonidae. They travel thousands of miles at sea on huge migration circuits which eventually bring them back to the coast close to the same river mouth from which they emerged as infants. The degree of return to their natal site for spawning is very high, prompting great interest in their means of navigation. But what of those few salmon that do not return to their birthplace? For many years it was presumed that these individuals somehow became lost due to a failure of the directional mechanism, so that they had to spawn elsewhere. However, recent research has shown that this is not the case. The salmon that "got lost" were those that had spent the longest time in the river system where they were born and had traveled the least distance from the coast once in the sea. On the other hand, a species such as the pink salmon, which spends only a few weeks in a river system after hatching and subsequently migrates up to 2400 miles from the coast, is characterized by a very high degree of return. Some 90 percent of those that survive at sea return to their birthplace. Other salmon which do not travel far are the salmon trout, *Salmo trutta*, which may never leave its river system, and the sea trout, which remains in coastal waters and shows a particularly low degree of return.

What is the explanation? It is now believed that those salmon species that spend all their lives within their river system or around the mouth of it have the most opportunity to explore suitable alternative spawning sites. Those species that leave the environment of their birthplace quickly to migrate over long distances have little opportunity to find other spawning grounds. To waste as little time and energy as possible at spawning time, they must use all the information at their disposal in order to return to their own birthplace, even if other suitable sites are available.

Mystery of the Spawning Salmon

How they actually find their way is slowly being revealed. It is easier to catch the salmon and perform experiments on them in river systems than in the open sea, so most studies have concentrated on the fish in their freshwater habitat. If salmon eggs or the newly hatched fry are moved from one river system to another, they invariably return as adults to the area where they developed into young fish. This is because they learn the location and characteristics of their "home" river during this period.

Experiments have shown that chemicals in the water and the alteration in their concentration at different points on the river's course are detected and memorized by the young fish. Fish have the ability to interpret different river "smells," and to them each river has its own probably unique olfactory signature. They can even distinguish between tributaries of the same river. Thus a salmon literally "smells" or tastes its way home to its birthplace by picking up a succession of familiar signs which point the way to the correct river system.

Another migration spectacle of the fish world is performed by the eels, a family of 16 species which belongs to the genus *Anguilla*. They spawn at sea and mature in fresh water, a movement described as *catadromous*. The eels spawn in the warm salt water of the western Atlantic at depths between 1300 and 2300 feet. The eggs hatch into leaflike larvae called *leptocephali* which then drift with the ocean currents toward the coasts of North America and Europe. The leptocephali then metamorphose into a more eellike shape just under 3 inches long to become *elvers*. These enter estuaries and rivers and gradually develop into eels. After several years the mature adults migrate back to the Sargasso Sea, even covering short distances overland if forced to by obstacles, in order to spawn, after which they die.

One surprising fact is that no adult eel has ever been caught beyond the waters of the continental shelf on its way back to the spawning area; they have only been found in river systems that drain directly into deep water. Adult eels have the enlarged eyes

Below: migrating salmon. The movements of salmon in the oceans can be studied by the use of various marking techniques. One of the most effective of these is the use of a small plastic tag, which is tied with silver wire through the muscles of the back of young fish during their descent to the sea. Once in the sea the salmon disperse in all directions. Such studies have been made on the salmon of the Pacific Ocean, when they are recaptured on long lines strung with thousands of hooks.

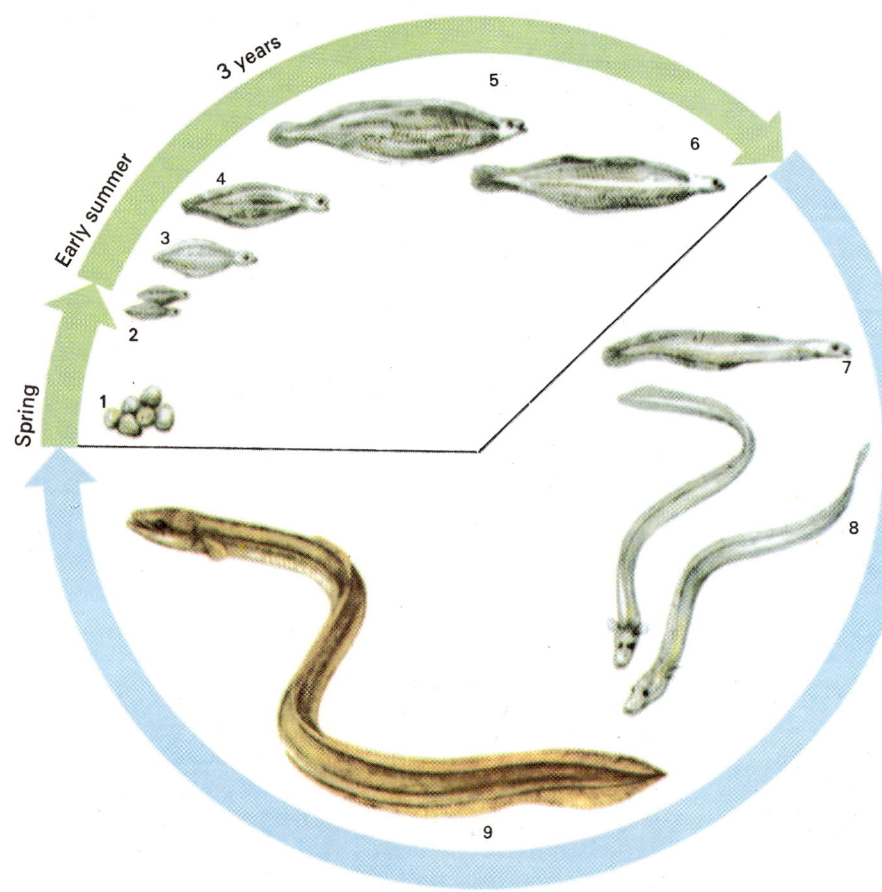

Right: diagram showing the life cycle of the European common eel. The adults—possibly only those of the American variety—spawn in spring in the Sargasso Sea (1) and then die. On hatching (2), the tiny eel larvae or *Leptocephali* grow and gradually change shape (3–6). As they develop they drift slowly with the currents of the Gulf Stream. In three years' time they reach European coastal waters. Here they change into transparent "glass eels" or elvers (7). By the time they have reached their destination—the mouth of a river—they have grown considerably (8). The adult eels (9) live in the rivers or estuaries, occasionally traveling overland by night when it is cool and damp. The males move out to the sea when they are from four to eight years old, but the females do not do so until they have reached seven to twelve years of age.

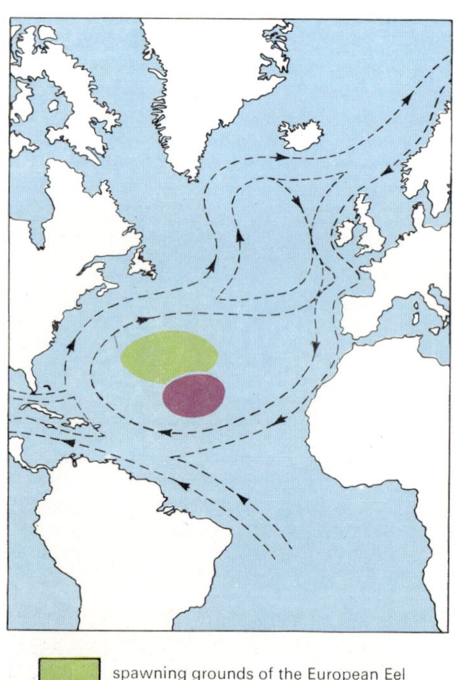

Below: map of the migrations and spawning areas of European and American eels.

and retinal pigment characteristic of deep-sea fishes, and they must use their huge eyes to navigate at depths so great that as yet none has been caught.

Most zoologists believe that the American and the European eel (also found in North Africa) are two different species. They are identical except for the number of their vertebrae. The American eel spawns from early spring to summer in the more southerly part of the Sargasso Sea, while the European, spawning over the same period, does so further north. The American leptocephali grow more quickly than the European, and by the time surface currents have carried them to the coasts, from the Guianas of South America to southwestern Greenland, they have metamorphosed into elvers. This only takes about a year. The European larvae have much further to travel, a journey which takes three to five years. Their rate of metamorphosis is timed by some unknown means to coincide with their arrival at the coasts of the eastern Atlantic Ocean and the Mediterranean, North, Baltic, and Black Seas.

One fish which is known to regulate its breeding and migratory behavior according to the phases of the moon is the 6-inch-long grunion of the family Atherinidae, which lives in the coastal waters of the Pacific off southern California. On those nights between March and August when the tides are highest (at the full and new moon) these silvery fish struggle as far up the beaches as possible. Their writhing bodies gleaming in the moonlight, the females wriggle up past the high water mark before turning

The Migration of the Grunion

to enter the sand tail first. The males coil themselves around the female's body to shed their sperm as the female releases her eggs. The grunion's biological clock is timed so that the eggs lie buried in the dry sand for the next 14 days, hatching when the next high tides wash over them. The young fish would die if they hatched earlier or later because they would not be able to reach the sea from the dry beach.

Research has shown that the eggs are fully developed after only seven days, but the hatching is only triggered off when the next high tide shakes the sand about 14 days later. Eggs developed in a tank in the laboratory, if subjected to a shaking after seven days, will hatch. It is still a mystery how the timing mechanism works in the adult grunion so that they know when the tides are at their highest.

The migration patterns of sea turtles, which are almost completely adapted to life at sea, are similar whether the species is the green turtle, loggerhead, hawkbill, Ridley, or leatherback. Definite migration circuits exist—best known is that of the green turtle, which travels from feeding grounds off the Brazilian coast to nesting beaches on Ascension Island in the mid-Atlantic. Turtles choose partners and mate at sea (where they are difficult to study), after which the females swim to the beaches where they will lay their eggs. Here they drag themselves out of the water and laboriously dig deep holes in the sand above the high water mark, in which they deposit the round white eggs, about 100 at a time. Seven to ten days later they hatch, and the hatchlings fight their way to the surface. They usually emerge at night to avoid the heat of the sun, and they immediately set off for the sea. Many

Below: a mass of writhing, slippery bodies of Pacific grunion spawn on a California beach. From March to August the grunion ride in on the surf of the extremely high tides that follow a new or a full moon. Each female is accompanied by one or more males who wrap themselves around her, fertilizing her eggs as she lays them.

Amazing Abilities

of the hatchlings do not make the water's edge because gulls, crabs, and other hungry predators find them easy prey.

The females, having laid their eggs, migrate to their permanent feeding areas where they spend the next two to three years, after which they return again to their nesting beach. Whether it is their natal beach is not as yet known, but having once chosen a beach the females show a high degree of return in future years although their navigation mechanism is as yet undetermined. With the large number of nesting sites throughout the tropical waters of the world it is a difficult task to track the females at sea, and as yet no turtle under a year old of either sex has been captured.

Certain marine mammals such as seals, sea lions, and walruses of the order Pinnepedia also return to nesting beaches to breed and give birth, while the cetaceans, comprising the whales, dolphins, and porpoises, are completely aquatic. Both groups show distinct migratory patterns in traveling from feeding to breeding sites.

Only a limited number of beaches are suitable breeding grounds for each species of seal, so great numbers of them gather there for a few weeks each year. Afterward they disperse and travel distances often of over 1000 miles, only to return the following year to the same beach. How they manage to find the right spot is still a mystery, although it has recently been established that most species can use sounds and their echoes as do dolphins, and this might help them to learn a route. However, the pups do not return for three years or so when they are mature, and even after that length of time they still locate the right beaches. Northern fur seals can travel over 6000 miles on the round trip from their breeding grounds on the Pribilof Islands off Alaska down the west coast of North America to California, or from the Komandorski Islands of Russia south to Japanese waters. How they navigate these immense journeys is not yet known.

Whales and their relatives are untiring travelers, many species

Below: young green turtles soon after hatching. More than a hundred eggs are laid by a female in one hole or nest about 2 feet deep. She then spreads sand over the site and, during the night or in the early morning, she abandons the nest and crawls back to the sea. The tracks left in the sand give the impression that a miniature tank has passed up and down the beach. Each female turtle lays five or six clutches of eggs at two-week intervals. The eggs hatch from seven to ten weeks after being laid in their underground nest. The young turtles, on emerging from the sand, head immediately for the sea. At this stage they are only about 2 inches long. Many newly hatched turtles are eaten on their journey to the sea by gulls, vultures, crows, and other predators as they are such easy game, but a few manage to reach the water. Little is known about the subsequent movements of the surviving young green turtles.

undertaking annual trips that cover several thousand miles. The whalebone or balleen whales—which include the largest living animal, the blue whale, which can be as long as 100 feet or more —form two distinct groups, one group inhabiting the waters of the northern hemisphere while the other remains in the southern oceans. These giants feed on the rich supply of plankton including a shrimplike crustacean called *krill* of the family Euphausiidae which lives mainly in the cold Arctic waters. Here the whales spend the summer feeding period consuming enormous quantities of krill which, stored as blubber or fat, must see them through the next few months. In autumn they migrate to tropical waters where they breed and give birth. They use the same routes year after year, a habit which has contributed to the decline in their numbers because whalers knew where to intercept them.

On these journeys the whales cannot be obtaining guidance from coastlines because they are usually far away from the nearest shore. It is known that they use sonar to navigate as do bats and sea lions, but it is not yet known how this would work over such great distances.

Whether on land, in the sea, or through the air, the navigational methods used by animals continue to astound human beings. Lacking an acute directional sense ourselves, it is particularly difficult for us to understand such abilities. A great deal of study is still necessary before we can truly claim to understand the mysteries of animal navigation.

Below: two humpback whales (*Megaptera nodosa*), each weighing about 40 tons, leaping from the water. Humpback whales have two distinctive physical features—furrows on their throats and very long front limbs. They migrate from their Antarctic winter quarters to various places including the east and west coasts of South America, southern Africa, and Australia, from which they then return to the Antarctic, often to the same place, the following summer. When on migration the whales tend to stay close to coastlines, a habit that has made the humpback very easy prey for whalers. While traveling they often leap clear of the water only to crash down on their backs, and the noise this splash makes can be heard several miles away.

Chapter 9
Secrets of Survival

The world is a hostile place for almost all living things, which have therefore developed some extraordinary techniques for insuring their own survival or that of their species. The ingenuity of nature makes for a fascinating survey—the lavish display of the male peacock or bird of paradise, calculated to attract the female and encourage reproduction; the frightening "owl eyes" adorning the wings of some butterflies, meant to scare off potential predators; the compelling fragrance of a flower, which lures flying insects and insures pollination; and the simple device of rapid reproduction, by which microscopic plankton guarantee survival while huge numbers of them are eaten by many water-based animals. The sometimes mysterious ways nature has evolved serve to protect living things from the specter of extinction.

Every shade and color can be found in nature, from the intense iridescent colors of a peacock's tail to the drab speckles of his mate, the peahen, among the birds, and from the brown twig caterpillar to the glamorous birdwing butterfly of the insect world. The color of a plant or animal is usually connected with survival value and how evolution over millions of years has equipped the organism for its life in a particular environment in relationship to other plants or animals.

One important section of the animal kingdom is composed of animals that are eaten by others. Their adaptations to protect themselves from animal (or plant) enemies are as diverse as those evolved by the predators to facilitate foraging for prey. Extremely small creatures such as microscopic plankton have little recourse and show few obvious preventive adaptations. Quick reproduction and its result—immense numbers—insure their survival, in spite of the fact that predators from ducks and flamingoes to fish and balleen whales consume millions of tons of plankton annually.

The larger the prey the more important color seems to become. Lacking the ability to flee or the defense of spines or tough armor, most animals disguise themselves by blending in with their background or looking distasteful or like a known enemy. As plants

Opposite: hornet moth (*Aegeria apiformis*), abundant in Europe but less common in North America. Its reddish brown and yellow bands give it a strong resemblance to a hornet, which it mimics in order to gain the hornet's safety from bird predators. Hornet moths are as large as hornets and have transparent wings. Their larvae live in the roots and lower trunks of poplar and willow trees and require two years to make their transformation into moths.

Above: cryptic coloration. The protective coloration of this woodcock makes its nest very difficult for a predator to find.

cannot move they have evolved the alternative adaptations of thorns or toxic chemicals to discourage herbivores from eating them.

The brilliant colors of the animal world are attractive to most people, probably because we ourselves are visually oriented and find color stimulating. But equally important are the duller shades which camouflage many different species and protect them from detection by predators. *Cryptic coloration* is achieved when an animal resembles the color or pattern of its usual surroundings, as with the woodcock, peahen, or peppered moth. Another form of *crypsis*, or visual protection, is called *disruptive coloration*. In this case the overall outline of an animal's body is obscured by a dramatic pattern, often of black and white markings—examples are the okapi, giant anteater, and zebra. A third cryptic technique is termed *countershading*, in which the animal's colors shade into each other in such a way that the shadows which give it three dimensions are eliminated and the creature looks flat, thus protecting it from those predators which use body shadows to locate their prey.

The caterpillar in its natural habitat clings upside down to a twig. Its back or *dorsal* side usually faces down, away from the sun, and under normal conditions would be in shadow. The *ventral* surface or underside normally receives the sunlight. The caterpillar is therefore lighter in color on its dorsal or shaded side than on its ventral side, which is comparatively dark. Usually it is the dorsal surface of an animal that is darker and the ventral that is lighter, as in mammalian species such as antelopes and gazelles, fishes including the rudd and the roach, and birds such as sparrows and thrushes. The effect of normal lighting is to make

Protective Coloration

the dark surface appear the same as the lighter but less well illuminated surface, with the result that the creature is more difficult to spot from a distance.

In experiments with detection the caterpillar larvae of different moth species—kentish glory, eyed hawk, privet hawk, and puss moth—were placed in both their normal, upside-down positions and abnormal positions, on the tops of their branches. Those deliberately placed on the tops of branches were first killed so that no movement would give them away. Three jays were then introduced, and the result was that they had far greater success locating the larvae on the tops of the branches than those hanging underneath. This confirmed that countershading in these caterpillars does afford some camouflage and protection from predators.

The circumstances of a particular predator-prey relationship determine which means of protection or combination of strategies will reduce to a minimum the prey's chances of being eaten. For example, in the short grass of the Serengeti plains in East Africa large hoofed animals such as zebras, gnus, and antelopes have nowhere to hide from their predators—lions and cheetahs—so they must rely on swiftness if both disruptive coloration and countershading fail.

Many animals which run the risk of being eaten have evolved a close resemblance to an inedible object. There are countless examples, but among the most obvious are the small insect in-

Left: disruptive coloration. This okapi (*Okapia johnstoni*) is protected by the fact that its body outline is obscured by its striking horizontal markings.

Mutants and Their Survival

vertebrates that look very much like sticks, flowers, leaves, and even bird droppings. The head, antennae, and legs of the creature are so well disguised that predators have great difficulty recognizing the living animal as prey. The American long-horned grasshopper called a katydid, of the subfamily Phaneropterinae, is a striking example. When resting, the legs are concealed either by being folded back upon themselves or protruding in a stiff, unleglike fashion. The antennae are usually laid along a branch or hidden in a groove of the extended forelegs. The mantis *Acanthops* can resemble a dead leaf and even conceals its head by partially hiding it under the folded front legs. The pupae of many insects are also well disguised—the pupa or chrysalis of the orange dog butterfly, *Papilio anchisiades*, of Trinidad hangs in a prominent position on a lichen-covered branch, but from above and side the resemblance to a small lichen-covered twig is perfect.

The peppered moth, *Biston betularia*, is probably the most frequently quoted example of evolution in action. This moth, which as a larva is a beautiful stick caterpillar, becomes in its adult form an excellently camouflaged insect. Like most moths it is nocturnal, flying only at dusk or at night. During the day it

Below: African bush cricket (*Preussia lobatipes*), which exactly resembles the leaf on which it nests.

Left: pale and *carbonaria* forms of the peppered moth (*Biston betularia*) at rest on a soot-blackened tree. H. B. Kettlewell performed a series of experiments on the peppered moth population around the Manchester area of England.

rests motionless on a tree trunk where the typical black, gray, and white peppered pattern of its body blends perfectly with lichen-covered trees.

A rare form of peppered moth before the Industrial Revolution was the mutant form *carbonaria*, which is almost completely black due to the black pigment melanin. With increasing industrial pollution, especially in northern England, the volume of soot and smoke pouring from factories darkened the trees nearby and killed off many lichens. By 1900 the *carbonaria* had become common in industrial areas and the pale form had become rare. In and around Manchester *carbonaria* accounted for 95 percent of the peppered moth population. In 1953 the British biologist H. B. Kettlewell performed experiments in which he released large numbers of both varieties of peppered moth, which he had marked under their wings, in two separate areas, one polluted and the other rural and more or less unaffected by pollution. Using this mark-release-recapture technique he found that the black form quickly disappeared from the clean, lichen-covered trees while the pale form disappeared from the industrially polluted wood. Kettlewell believed that each type disappeared from trees of a contrasting color because predators could find them more easily. He filmed the predators in action with the help of Nikolaas Tinbergen, a pioneer Dutch ethologist or animal behaviorist. They showed that thrushes, flycatchers, yellow hammers, robins, song thrushes, and nutcatchers in the clean wood claimed a predominance of *carbonaria* moths, while in the polluted wood the main predator, the redstart, took three times as many typical peppered moths as it did *carbonarias*.

Further laboratory and field work showed that from an original proportion of just one in 1000 the black moth form became dominant, to the extent of making up 95 percent of the population in an industrial habitat. This was largely due to the different degrees of camouflage achieved and the effects of natural selection. Predatory birds tend to eliminate 30 percent more of the

conspicuous form than of the camouflaged one, and which is which depends on the habitat.

In Britain research has shown that over 100 moth species have changed in this way. The phenomenon, known as "industrial melanism," is also widespread throughout North America and Europe. However, there are now signs that the population ratios of black to peppered moths are returning to the 1850 situation when the black form was rare. This is due to the introduction of cleaner industrial practices, smokeless zones, and the resulting gradual return of clean trees that can support the growth of lichens.

Even if an animal is well camouflaged, however, it will not survive for long unless its behavior corresponds to its cryptic coloration. If an insect which mimics a leaf rests on a leaf-free tree trunk, or if a stick insect moves rapidly along a branch, it will not fool a predator for long. So camouflage involves not only body shape and color pattern but also the animal's behavior pattern, which may also be characterized by deception. For example, katydids and mantids move very slowly when they have to move, or they may only rock slowly from side to side like a leaf in a slight breeze. Rapid movement would attract the attention of a predator, which might come closer to investigate.

Other insects, such as the garden tiger moth, *Arctia caja*, employ a dash-and-freeze movement when in danger. When the moth takes off from its resting place it displays a flash of bright color, which confuses the attacker as the moth becomes conspicuous one second and disappears the next. The predator probably fixes on the place where the flash occurred, thus losing track

Above: bull's eye or io moth (*Automeris io*), whose hindwings possess dramatic eye spots or ocelli which can be covered by the movement of its forewings. In experiments it has been shown that birds draw back when the "eye" suddenly appears.

Opposite: American long-horned grasshopper or bush cricket known as a katydid, loudest of the grasshoppers. They feed mainly on leaves and their chief enemies are birds.

of the prey. Some cryptically colored tropical frogs are brightly colored on their inside hind legs. If disturbed by a potential predator they jump, exposing the color, then quickly freeze and become hidden once more.

The flash coloration is taken a stage further by the eyed hawk moth of the family Sphingidae; the atlas moth, *Attacus atlas*; and the peacock butterfly, *Nymphalis io*. On their hindwings, hidden from view when the insect is at rest, are a pair of huge *ocelli*, dramatic spots which look like the large eyes of an owl or other bird of prey. By moving the upper pair of wings the lower wings are exposed, and the huge eyes frighten off small predatory birds. The caterpillar of the hawk moth also has its own protective behavior. If disturbed, it blows up its head and thorax until it looks like a small poisonous snake, and the elephant hawk moth caterpillar, *Deilephila elpenor*, even has a false pair of large shiny eyes to increase the effect when it rears up and weaves back and forth, hissing like a snake.

Most cryptically protected creatures rest, camouflaged, during the day and move around to feed themselves during the hours of darkness when they are less likely to be detected by predators. But many animals have no camouflage—on the contrary, they exhibit conspicuous and dramatic colors, evolved to warn potential attackers that they are distasteful or poisonous. The brilliant

Below: elephant hawk moth caterpillar (*Deilephila elpenor*), which shows its ocelli when molested.

red-and-black, yellow-and-black, or green-and-black markings of the various species of South American arrow-poison frogs, of the genera *Sminthillus*, *Dendrobates*, and *Phyllobates*, serve to advertise to potential predators that they are dangerous to eat. The first time an attacker comes across the frog it may touch and perhaps pick it up in its jaws. However, on tasting the frog's poisonous skin secretions, the aggressor quickly drops its meal and thereafter associates the unpleasant experience with the frog's striking color patterns, and in future it will avoid them. The same method works with the bright orange-and-black markings of the monarch butterfly and the red-and-black markings of the cinnabar moth—they are both distasteful if eaten by a bird due to the high concentration of histamine in their bodies.

Many other insects show warning colors to advertise their presence, secure in the fact that they are not worth eating. Red-spotted burnet moths, *Zygaena*, discharge prussic acid (hydrogen cyanide), and the familiar ladybirds or ladybugs, with their black-and-red or black-and-yellow markings, are also poisonous. Unfortunately for them, however, there is usually some creature that will feed on them regardless, either because it is immune or otherwise unconcerned by the deterrent. The cuckoo, for example, readily attacks poisonous caterpillars that other birds soon learn to leave alone.

Danger Warning!

Left: red and black arrow-poison frog of the genus *Dendrobates*, found only in Central and South America, where the Indians extract poison from their bodies to use on arrowheads. They do this by piercing the frog with a sharp stick and holding it over a fire. The heat forces the poison through the skin where it collects in droplets, which are then scraped off into a jar. The venom of the arrow-poison frog is now being used in the laboratory for studies on the nervous system. It has been found that it acts as do the hormones secreted by the adrenal gland, which block the transmission of messages between the nerves and the muscles. Large amounts rapidly cause death, but in tiny doses the poison may be proved to have medicinal value for humans.

The Meaning of Mimicry

Below: Batesian mimicry in butterflies. The male diadem butterfly (top) looks like many other members of its nonpoisonous genus *Hypolimnas* of the family Nymphalidae. However, the female diadem (bottom) resembles the plain tiger butterfly (middle), belonging to the Danaidae family which is inedible to birds. The edible female diadem thus gains protection from the resemblance.
Below right: drawing by Bates of a bird-killing spider that attacked finches, published in *Naturalist on the River Amazon* in 1863.

Opposite: drawings by Bates of some of the insects he observed, each meticulously colored and carefully numbered in sequence.

Some creatures, neither poisonous nor distasteful, have evolved the same color patterns as animals which *are* protected by defense mechanisms. This relationship is termed *Batesian mimicry*, named after its discoverer Henry Bates, a 19th-century British naturalist famous for his 1848 journey into the Amazon basin of South America. He discovered many cases in which palatable insects had forsaken the cryptic patterns of close relatives and had evolved bright colors resembling those of distasteful species, and Bates realized that these insects were successfully tricking predators into avoiding them.

Numerous examples of Batesian mimicry can be found in the insect world. Bees, wasps, and hornets are mimicked by numerous bugs, moths, beetles, and flies which have no sting. Some ants are mimicked by spiders, some ladybirds by cockroaches, and parasitic ichneumon wasps by cockroaches, bugs, and flies. There are also examples among the vertebrates—in Africa the crested rat mimics the zorille, an animal with the scent glands and black-and-white warning coloration of the American skunk. If attacked the rat raises its hair to expose the familiar black and white stripes.

Predators have also been known to mimic harmless animals in order to approach their prey more easily. This behavior, like that of the wolf in sheep's clothing, is exemplified in the zone-tailed hawk. It closely resembles the turkey vulture or buzzard of North and South America, not only in form but in behavior. The hawk frequently flies amongst vultures, which because they do not usually hunt live animals do not present a threat to any creatures below. Thus the zone-tailed hawk gains the advantage of surprise when it drops down from the sky to grasp an unsuspecting rabbit or other small mammal.

No 1
No 2
No 3
No 4
No 5
No 6
No 7
No 8
No 9
♂
No 10
No 11
No 12
No 13
♀
No 14
No 15
No 16
No 17
No 18
♀
No 19
No 20
No 21
♀
No 22
No 23
No 24

Displaying Their Finest Feathers

Mullerian mimicry is another form, named after the 19th-century German zoologist Fritz Müller who lived for some time in Brazil. After collecting many distasteful butterflies he noticed that different species were often marked with extremely similar warning colors and patterns. He pointed out that this was an advantage to all of the species because potential predators would only have to learn to avoid one warning pattern instead of many. It follows that the first attacks will be shared out over the whole group, and each species will suffer only a few losses on average. This is the case with the passionflower butterflies of Central and South America. When so many species look alike, each individual must find an appropriate mate by smell. Their different scents are apparent even to the poorly discriminating human nose. Sometimes the same warning patterns are found on totally unrelated animals—for example, the bright black-and-yellow warning colors of the poisonous caterpillar of the cinnabar moth, and of stinging wasps and bees. It may even be possible that a predator, having once been stung by a wasp, avoids not only the cinnabar caterpillar but also any other black-and-yellow animal such as some types of salamander.

In addition to situations of predator and prey, there are thousands of examples of strikingly colored but nonpoisonous creatures which do not mimic other animals. Their coloring is often used for social signaling within the species. Examples are the vivid red-and-blue face of the mandrill, an African primate; the bright feathers of the birds of paradise of New Guinea and the

Right: the bright colors of a male mandrill (*Mandrill sphinx*), perhaps used for social signaling or to confuse its enemies. Primates have better color vision than most mammals.

Above: marine butterfly and damselfish swim and shelter in the coral (*Pavona*) of a tropical sea. A coral reef is an exotic world. Its population is varied, colorful, and closely interrelated.

tiny hummingbirds of the New World; the colorful folds on the throat of the anole lizard of the Americas; and the denizens of tropical coral reefs such as the jewel and butterfly fishes. The majority of these animals are active by day when their colors can be seen, and they also have good color vision themselves: the range of colors visible to many animals is different to that visible to humans. Bees and butterflies have the ability to see the ultra-violet shades of the flowers on which they feed. Birds apparently cannot distinguish different shades of blue, while insects are colorblind for red.

Because the majority of birds are excellent fliers and so can escape any dangerous intruder by simply flying away, many can afford the conspicuousness of brilliant and dazzling colors in their wing and body feathers. The most colorful birds are found in tropical jungles, where they can disappear from sight among the shadows or even in the open, since they could be taken for bright tropical flowers at a distance.

The bright colors and patterns of birds serve not only for attracting a mate but also for aggressive displays. The iridescent, "eyed" train of a peacock is used mainly in courtship—the male fans it out so that it shimmers as he performs around the female peahen. In contrast, she is a drab brown for camouflage on the nest. Some game birds, on the other hand, such as the grouse and prairie chicken, use their colors as part of their threatening behavior in staking out a territory as well as for attracting the opposite sex. The peacock's relatives the pheasants, members of the duck family, and some perching birds such as lyrebirds and birds of paradise, are also equipped with special means of display.

Above: male red jungle fowl (*Gallus gallus*), best known of the four species of jungle fowl and thought to be the ancestor of the domestic chicken. The cocks have the characteristic fleshy appendages on their heads—a comb rises from the top of the head and lappets hang under the chin. The cock's plumage is mainly red and black with an iridescent greenish sheen, his bright colors being used for attracting a mate and aggressive display when claiming a territory. They live in forests in the warmer parts of Asia from India to southern China and Indonesia. Populations were established on various Pacific islands and in South Africa by human settlers from Europe.

The birds of paradise of the family Paradisaeidae are the most ornate and colorful assemblage of birds in the world, although only the males possess these special feathers for attracting the female. Few observations have been made of these birds in the wild because the 43 species live in the rain forests of New Guinea or the mountain forests of northeastern Australia. Much of what is known is from studying captive birds, and many aspects of their lives have yet to be learned. Raggiana's bird of paradise is one species that has been studied in the wild. During the breeding season, either at dawn or in the early afternoon, the male calls to females to come and watch him perform from nearby trees. He starts his courtship dance by hopping sideways on a branch, then he opens his wings to reveal a fine array of red feathers. He then arches forward until his head points to the ground, which allows his tail feathers to cascade into a shimmering display. He repeats his performance until a female is attracted to join him on his perch, after which mating takes place.

Few mammals are brilliantly colored, because the majority see only in shades of monochrome. The primates, however, are characterized by good color vision so it is among them that the more striking mammals are found. The award for the most impressive face must go to the male mandrill, a forest-living baboon of Africa. Although the rest of the body of this baboon is an ordinary dark brown, its face is distinctly different—the nose is bright red with dazzling sky blue ridges on the muzzle. He even presents a striking picture as he turns away—his genitals are the same colors and the hairless rump patch is blue merging into the violet and pink of the sitting pads. Zoologists have found it difficult to explain the purpose of these brilliant colors. One suggestion, put forward by the German ethologist (animal behaviorist) Wolfgang Wickler, is that the mandrill is highly colored both front and back in order to baffle his enemies, which include the leopard. Desmond Morris, the British ethologist, does not accept that an enemy would be confused in this way, but he suggests that two colorful areas could make a double impact in a threat display, whether to another mandrill or a potential enemy. A perplexing point is that the drill, a closely related species, has a black face but retains the colorful rump of the mandrill. Whatever the purpose of these colors, both species seem to flourish in the wild, and perhaps further study will reveal the answer.

Perhaps the most colorful baby in the mammalian world is that of the spectacled langur, a small slender monkey of the forests of southeast Asia. The parents have blackish coats and a contrasting white ring of naked skin around their eyes. The babies are born with a striking orangey-red coat. The other 19 langur species produce babies which are also different in color, usually yellow or white, from their parents. The reason appears to be involved with methods of caring for the infant: the mother feeds the baby but other female "aunts" take turns looking after it, passing it around among themselves. The bright color, which changes to the darker adult color at the age of three to five months, probably helps the mother locate the baby if it is in distress.

Color is also found in the world of reptiles for purposes of courtship and threat display. The male anole lizard, basically

The Colorful Courting Fish

well camouflaged in green and brown, has the facility for expanding the dewlap or gular pouch beneath his throat into a yellow organ of display. He does this, sometimes even bobbing up and down simultaneously, in order to defend his territory against another male anole. The display is also used to attract females. Each species has its own rhythm and type of pouch so that individuals perform only to members of their own species.

Many fishes, like birds, can afford to be brightly colored because they can easily escape from their enemies. Many fishes that live in coral reefs, such as parrot fish, butterfly fish, and angelfish, show a vivid range of colors and patterns. Their strikingly marked bodies are for species recognition and display purposes. They do not need camouflage to stalk their prey because they feed on

Left: eggs in the nest of the three-spined stickleback fish (*Gasterosteus aculeatus*), whose courtship ritual was studied and photographed by Nikolaas Tinbergen. The male stickleback changes color during the spring breeding season in order to attract a mate to his nest.

stationary organisms such as worms, coral polyps, and algae. If predators approach they can simply retreat into the safety of cracks and crevices in the coral, where they also breed.

Nikolaas Tinbergen studied the role of color in the courtship behavior of the three-spined stickleback, *Gasterosteus aculeatus*, in the 1950s. These small freshwater fish swim in small shoals for most of the year, but during the spring breeding season each male isolates himself to establish his own territory. In the process he changes his appearance: his eyes become shiny blue, his dull brown back becomes iridescent green, and his belly is suffused with bright red. He laboriously builds a nest by first scooping a hole in the sand then using secretions from his kidneys to glue together filaments of weed and algae. Finally he performs a particular sequence of movements to attract females into the nest.

The different animal species of the world—all 1,100,000 or more of them—are specifically adapted for the lives they lead in their own particular habitats. Each species has its own survival techniques which insure that sufficient members will mate and enough offspring reach maturity to guarantee the survival of the species for generation after generation. The ingenuity of some of the methods is amazing, but whatever the tactics they usually work only in the right environment. Although many are killed through disease or starvation, and others are eaten by predators who in turn rely on this food for their own survival, enough usually survive, even in their hostile world, to guarantee that the species survives.

Chapter 10 Animal Architects

Why do animals build? Obviously for shelter—to provide safety from predators and bad weather, for food storage, and as a place to give birth to their young. Some encase themselves in a silken material produced from their own bodies, and inside the safety of these structures the early larval stage develops into adulthood. The need for shelter is the same whether the animal builds it or carries it around as a specially adapted part of its body. Scientists have always been curious about why each species builds its own characteristic home, and they have taken a particular interest in the techniques of bees, ants, and termites, insects which live in large communities and whose architecture is extremely precise. But much more painstaking research is needed before we come to understand fully the intricate mechanisms of animal architects.

In a world dominated by man we tend to think that only humans can build skyscrapers, huge apartment blocks, dams and reservoirs, bridges, and homes. In fact many animals surpass man in their capability to build and perform feats of engineering. Some can weave, others stitch, many use mud and plant materials to cement, some spin silk, and some build huge townships above and below the ground. Many human architectural accomplishments have been modeled on those of animals, and Sir Hugh Casson, the modern British architect, freely admits that birds and other animals are better architects than the human race.

Why do animals design homes? The answer is simply that the provision of safe shelter is a necessity for survival. The millions of soft-bodied coral polyps, small sedentary organisms that build the beautifully impressive coral reefs of shallow tropical seas, are protected by the calcareous or chalky shells they secrete which envelop their bodies. Caddisfly larvae of the insect order *Trichoptera* live in freshwater pools, lakes, rivers, and streams, where they construct tubular cases to protect their bodies. They choose their building materials depending on what is available in their particular habitat. Those that inhabit stagnant water make use of bits of reeds and leaves, while those in faster-moving water tend to use pebbles and shells to weigh themselves down so that

Opposite: an air-filled underwater balloon, home of the water spider (*Argyroneta aquatica*). Found in the stagnant waters of ditches and ponds throughout Europe and Asia, this spider lives its entire life underwater and yet remains an air-breathing animal by carrying its own supply of air around with it. The air surrounds the spider's abdomen and lower thorax like a silvery cloak and adheres tightly to the furry hairs of its body. To keep the air supply renewed it need only rise to the surface once a day. Its balloon home is held in place, between aquatic plants and submerged branches, by a network of threads spun before it is filled with air. To do this the spider rises to the surface, traps an air bubble by crossing its hind legs over its upturned abdomen, and carries the bubble down to its nest. It then releases the bubble under the web. By repeating this motion a dozen or so times, the spider can fill a hemispherical balloon of about an inch in diameter.

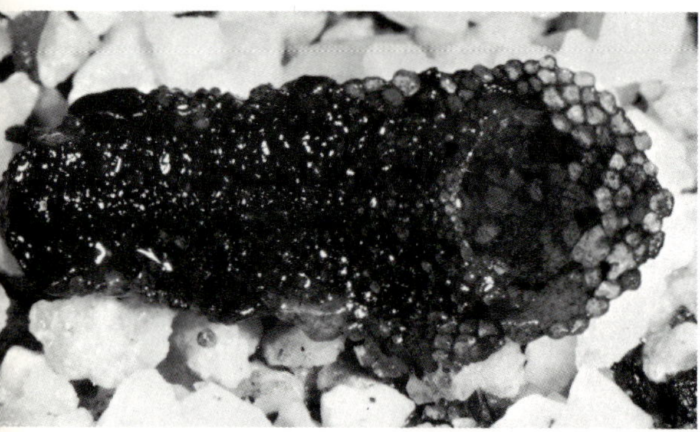

Above: caddisfly larva in its woven silken tube, with pebbles attached to weigh down the structure. The larvae pick up this heavier material using their legs and sharp-edged jaws, then cut it to shape the nest.

Below: the "cuckoo spit" of the spittlebug (*Philaenus spumarius*). When the larva is encased in its protective ball of froth ants and other predators usually leave it alone.

Below right: soft-bodied coral polyps, builders of the coral reefs of shallow tropical seas.

they are not swept away in the current. All of them bind their cases with silk they spin themselves.

To make its own personal case a larva, or "worm" as it is also called, weaves a little silken tube with material supplied by its salivary gland. It picks up with its legs and sharp-edged jaws plant material or stones from the surroundings to make the tube stronger. Its soft abdomen remains inside the tube at all times. Each species has its own architectural style, and some insect collectors specialize in collecting species of caddisfly larvae the way a lepidopterist collects and displays butterflies and moths.

Another larva that hides inside a structure it makes itself is that of a froghopper insect—the spittlebug, *Philaenus spumarius*. The larva hatches from an egg which is usually laid in the angle formed by a leaf and the stem of a plant. It then sucks sap from the plant so quickly that the liquid soon emerges through its anus. There it is mixed with a soapy fluid from glands on the underside of the abdomen so that the larva literally sits in a small pool of plant juice, whipping it into froth by expelling stale air into the fluid through its windpipe to form the "spit" familiar to gardeners. It is still a mystery, however, how the spittlebug manages to make the bubbles last, since froth in water usually disappears very quickly. It may be that some sort of secretion from its kidneys helps to stabilize the foam.

Foam nests are found not only in the world of insects but are

Chalky Shells and Foam Nests

quite often used by frogs, especially in tropical habitats. The foam is produced not by the developing tadpoles but by the parents to give protection to their young. Frogs that build foam nests include the Javanese flying frog *Rhacophorus* and the African tree frog *Hylambates maculatus*. A mated pair of Javanese flying frogs performs a typical amphibian mating hold, or *amplexus*, in which the male climbs onto the back of the larger female and grasps her around her neck or under her arms. The female chooses a large leaf or group of smaller ones on a branch overhanging a pool or stream and lays her eggs, each with a mucous covering, one at a time. The male fertilizes them as they are laid, at the same time creating a foam by dipping his feet into the mucus and kicking his feet together rapidly. This process is repeated until 60 to 90 eggs have been laid and fertilized and the foam has been whipped up into a ball some 2 to 3 inches in diameter. The female presses the surrounding leaves together over the eggs, and the foam quickly hardens at the point of contact. The parents then depart, leaving the tadpoles that hatch from the eggs in a mini-aquarium, as some of the foam liquifies to form a pool inside the ball. Their home is safe until the next torrential tropical rainfall dissolves the outer layer of dried foam and washes the tadpoles down into the fast-moving water current below. African tree frogs build very similar nests, but in theirs the foam is not allowed to dry up. The female waters the nests of

Below: a prickly cockle (*Cardium echinata*), one of about 200 species of heart-shaped, globular, bivalve (two-shelled) mollusks. The prickly cockle has spines along the ribs of its shell, is smaller than many other similar cockles, and is found widely over the globe.

A Nest of Bubbles

Right: sponge of the class Hexactinellida, which forms colonies underwater. A fresh sponge is slimy to the touch. Its soft body consists of innumerable cells penetrating and enveloping the elastic "spongy" skeleton made of a substance called spongine, consisting of proteins. The type of sponge shown here constructs its skeleton from siliceous substances.

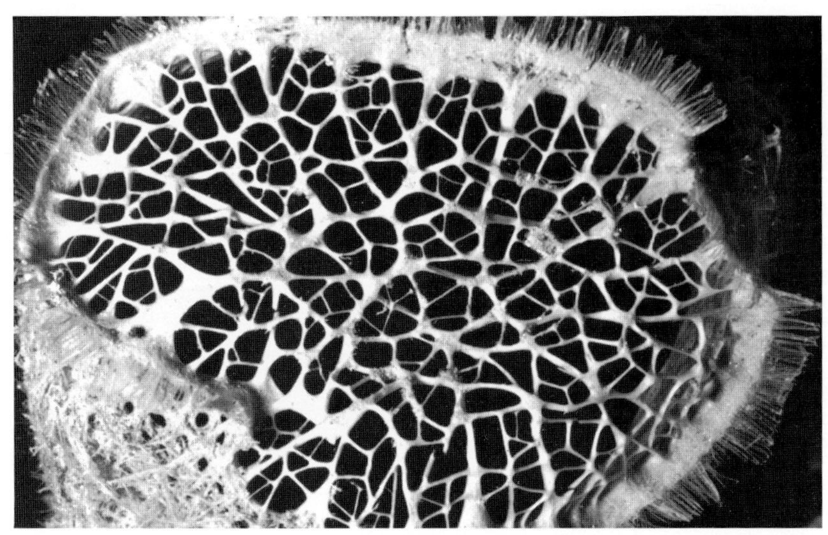

foam continually, and when the nest eventually disintegrates the tadpoles drop into the water below.

Several species of fish are also foam builders, having evolved bubble or foam nests to protect their young. The most famous are the fighting fish of Thailand and Indonesia (*Betta*) and their close relatives the paradise fish of southeast Asia (*Macropodus*), both of the family Anabantidae. They are *labyrinth* fishes, which means that they have air-breathing organs in addition to the usual gills for extracting oxygen from water. At breeding time the male selects a suitable site for the nest, then takes in bubbles of air at the surface which he releases underwater to form a raft of air bubbles. A mucus is produced by a hormone in the cells of the gill lining, and this is mixed with the air in the fish's mouth before it is released, thus stabilizing the bubbles. A female is enticed to mate beneath the nest, after which the eggs of the paradise fish rise automatically into the bubble nest because their yolks contain a large globule of oil, making them lighter than water. The male fighting fish, however, must catch each fertilized egg as it is laid, before it sinks, and spit it into his nest before resuming his mating with the female. When all the eggs have been laid and trapped in the bubble nest the female is chased away and the male guards them until they hatch some 24 to 48 hours later. The male paradise fish also cares for the fry after they have hatched, and if any strays too far he will suck it into his mouth, transport it back to the nest, and spit it out there.

The need for shelter is the same whether an animal builds it or carries it around, having adapted some part of its body to fill this requirement. Most terrapins, turtles, and tortoises have thick, bony shells into which they can withdraw their soft, vulnerable legs and head. The majority of snails hide their soft body parts inside a calcareous shell which they secrete as they mature. The armadillos of the southern United States and South America are protected from attack by tough plates, and some species can roll up into a tight ball.

However, most animals must rely on their instincts to prompt them to construct suitable shelter. Some, such as the kangaroo rats of the deserts of western North America, need shelter from intense heat. They excavate long burrows to reach cooler levels

Left: trapdoor spider (*Pachylomerns nidicolens*) of Jamaica, which inhabits sandy tropical regions. It digs a tube-shaped sloping burrow with the aid of comblike spines on the edges of its jaws. Then it coats the hole with a mixture of soil particles and saliva and lines that with silk threads. It weaves a strong lid to cover the nest, which is fastened to the top edge of the opening by a strong silken hinge so that it can be easily opened but will close automatically.

Below: a spider's web, specially equipped to catch flies. The thread is covered with a sticky substance to prevent the trapped fly from escaping. The more it struggles, the more entrapped it becomes. The spider then moves in quickly, paralyzes the fly, and eats it at its leisure.

underground. The ground temperature may be as high as 110°F but at the end of the burrow it is a cooler 80°F and the air is much moister, although the burrow is only about 20 inches below the surface. The jerboas or desert rats of North Africa build more sophisticated burrows, usually sealing the entrances to their underground retreats with sand to exclude hot air and leave no visible trace for predators. They use their whole body as a digging tool—the hands and hind legs work together in rapid movements to brush soil away, and the head and nose work as a living hammer to firmly pack the tunnel's soil wall. They dig deeper if the tunnel's temperature rises by too much, and in winter they line their nest chambers for warmth with shredded wool and any available vegetation.

Many rodents that burrow underground use their tunnel networks to store food. The wild common hamster of Europe and Russia collects cereal seeds, roots, and parts of green plants. It may accumulate up to 200 pounds of food which it carries back to the storage chambers of its burrow in its cheek pouches.

Some animals excavate underground nests for protection from cold and snow. In autumn male polar bears usually move southward from their summer Arctic home to seek out warmer weather and increased food supplies. At the onset of winter, however, the female digs a den in deep snow and in mid-winter she gives birth, usually producing two or three cubs weighing less than half a pound each. They are blind and almost naked, totally dependent on their fasting mother for warmth, food, and protection. In spring the mother cuts through several feet of winter snow to make an exit from the den.

Like the polar bear, many other mammals give birth to their young in the safety of an underground nesting chamber. They include foxes, moles, badgers, and brown bears, as well as the primitive, egg-laying duckbilled platypus, a monotreme. The platypus lives near rivers in eastern Australia, where it excavates a comparatively complicated underground nest which is remarkable in that little soil is ejected. It packs soft soil into the sides of a tunnel to form an arched roof and flat floor. The burrow terminates in a nesting chamber which is usually 15 to 25 feet from the entrance but can be as far away as 100 feet. The

Above: typographer bark beetle (*Typographus coleoptera*), one of 2000 species found throughout the world. Both larvae and adults feed on the bark or wood of trees. Usually they make tunnels in the bark itself, but quite often the tunnel is through the layers where bark meets wood. Each group of bark beetles forms its own distinctive pattern. The mother beetle usually makes the central tunnel and lays her eggs along the sides. When they hatch the grubs burrow outward.

platypus industriously lines the nest with moist grasses and eucalyptus leaves to keep the two eggs she lays from drying out.

Without doubt the master mammalian tunneler is the mole, an inhabitant of the northern hemisphere, which has "shovels" for feet, a sturdy compact body, and sensitive snout and feelers. It comes to the surface so rarely that its eyes have degenerated to the point of being almost useless. The molehill and surface runs (ridges formed when the mole drives its way along, less than an inch below the surface) are the only visible signs of this subterranean architect at work. Solitary except during the breeding season, the mole excavates tunnels at various levels depending on the season. The lower ones, which are below the frost line, are used in winter. The molehill itself is a vertical tunnel leading off a main run, the result of pushing unneeded soil up to the surface.

Other superb excavators are found among insects, especially the larvae of some beetles, moths, butterflies, weevils, and glowworms. The adult mole cricket of the family Gryllotalpidae is so called because its toothed and flattened forelegs, excellent adaptations for tunneling, enable it to plow through the soil in a similar fashion to the mole.

A large number of animals are adapted for boring or chiseling into wood rather than burrowing into the ground. Many beetles excavate tunnels—the characteristic patterns of bark beetles such as *Scolytus destructor* can be found on the underside of pieces of tree bark.

Woodpeckers of the family Picidae are the carpenters of the avian world. Their woodworking equipment consists of strong claws and stiff tail feathers which give the birds secure footholds and support as they move about on vertical tree trunks. The most important tool is their strong beak which is used to chisel out a breeding cavity in the wood of a sound tree. This hard work is done mainly by the male, and once constructed the nest will often be reused in successive years. The woodpecker's skull bones are especially strong in order to protect the brain from the vibration

Home Builders of the Bird World

of the continual hammering. The beak is in continuous use—apart from nest site excavation work it is used to find insect larvae buried in tree trunks, to open nuts and conifer heads, and to make drumming sounds to establish a territory and attract a mate.

A different kind of hole-breeder is the bizarre-looking, big-beaked hornbill of the family Bucerotidae, found in sub-Saharan Africa and tropical Asia. The beak in most species is not as heavy as it looks, because it contains many air-filled chambers. It is no good for chiseling but the sharp point can be used to spear fruits dangling from the tips of branches. The hornbill's unique nesting behavior begins when the mated female enters the nest hole, often a disused cavity in a tree, and walls herself in. She uses saliva and mud, clay or dung, which are collected for her by her mate. She hammers the mud into shape by vibrating her broad beak sideways, and when dry it is very hard. Though she is now safe from enemies she is unable to feed herself, so the hornbill male brings her fruit, berries, and even flowers. Once the young have hatched after four to eight weeks the female may break out of the nest to help the male find enough food to feed the growing chicks, or she may remain in the nest until they are ready to fly after another six to seven weeks. If the mother does leave the half-grown chicks they instinctively rebuild the wall, which they then break down themselves when fully feathered and ready to leave the nest.

The majority of animal builders construct their homes by adding materials they have actively sought out and chosen. The most notable are found in the avian world. Most birds need a permanent nest site for the breeding season because their future offspring all spend some time developing inside a fragile eggshell, and they need warmth and continual protection from predators. Nests vary greatly, but each suits the particular species. Some birds do not make a nest—emperor and king penguins act as living incubators of their single eggs by placing them on their large webbed feet enveloped by a large fold of the lower abdomen

Below left: the simple round nest of a chaffinch, one of the most artistic structures of the avian world. It is built by the hen of grass, rootlets, and moss, lined with feathers, and decorated on the outside with pieces of lichen tied on with cobwebs.
Below: king penguin (*Aptenodytes patagonica*) incubating the large single egg on his feet where it is protected by a fold of abdominal skin. The chick is also kept warm in this way until it develops enough fat to insulate itself.

How Did Nest Building Begin?

Right: cliff-dwelling seabirds called kittiwakes (*Rissa tridactyla*). They nest in large colonies, sometimes up to 100,000 pairs, on the narrowest of cliff ledges. Each kittiwake, unlike most gulls, builds a solid cup-shaped nest of mud, grass, or seaweed to prevent the eggs from toppling over the edge of the cliff. Even when the chicks have hatched they are unable to run around and must stay in the nest for the first few weeks, but because of the inaccessible position of their nests they are not in much danger from predators.

Opposite: common village weaver bird (*Ploceus cucullatus*), which displays great skill in nest building. Sparrow-sized birds about 6 inches long, they have the conical, seed-eating bills of the house sparrows. The common village weaver is golden yellow with a black head and black streaks on the back and wings. A single palm tree can hold hundreds of nests which hang like pendulous fruits from the branches. The birds strip so much leaf material from the palm fronds to construct their nests that eventually the palm will probably be killed.

to keep the eggs off the freezing Antarctic surface during the long incubation period. On hatching the baby sits on the feet of one of the parents until it is sufficiently developed to cope with the adverse climatic conditions.

The fairy tern or white noddy, *Gygis alba*, is a seabird found near most tropical and subtropical seas. Sometimes the female will not build a nest but instead finds a small hollow or suitable fork in a tree or rock and lays her single egg in it. The young baby tern has strong, curved claws which grip the tree and prevent it from falling off. The eggs of guillemots and many other cliff-dwelling seabirds are protected from rolling off the ledges where they are incubated by their conical shape—they can roll around without necessarily rolling over the edge.

Some female ground-nesting birds simply lower their bodies into soft earth and rotate them to form a shallow depression. The common tern of the family Laridae does this on sandy ground. Others such as ducks and geese do not gather material for their nests but mold the grassy vegetation into a nest. The female then adds an insulating layer made from soft down feathers plucked from her breast.

The origins of nest-building techniques remain a mystery. They are probably connected with behavior patterns in the same way as are many birds' ritual displays of grass pulling. A typical nest is a neatly woven cup of sticks and grasses with a soft lining, and the majority of perching birds—including the thrush, starling, linnet, and chaffinch—construct such nests. The material chosen is fairly pliable and the bird sits in the structure as it builds. Material is pulled and tucked into a framework of living branches and twigs, and gradually the rounded nest takes shape. Soft lining material is then added, at which point the bird rotates itself to mold a personal fit. The bird sits snugly in the final structure with room for the eggs beneath it.

The tiniest, neatest, most delicate nest is that of the hummingbird, over 300 species of which live in the New World. The female undertakes the duty of building the deeply cupped structure from

How to Weave Roofed Nests

Below: nest of the house martin (*Delichon urbica*). This bird originally nested on cliffs, and still does in Japan, but now it builds its nest of mud and saliva under the eaves of houses. Gathering the mud is a communal event, after which mouthfuls of mud are stuck together and strengthened with grass to form the typical round nest. The final result looks as if it is made of round pebbles glued together. Careful maneuvering is necessary for the house martin to land in its nest.

cotton, moss, lichen, and spiders' webs. At the other extreme is the clumsy assemblage of twigs and sticks amassed by pigeons. Frequently more twigs end up on the ground than as part of the pigeon's nest.

The range of birds' nests shows a great ingenuity and variety in construction. A modification of the woven, cup-shaped type of nest is one which includes a roof to protect the inmates from rain and other inclement weather as well as providing better cover from sharp-eyed predators. The magpie of the genus *Pica*, found in Europe, Asia, and North America, builds a twig roof, though it is quite thin and not completely waterproof. The tiny wrens of the family Troglodytidae, warblers of the family Muscicapidae, and longtailed tit or chickadee (*Aegithalos caudatus*) are Eurasian and American *passerines* or perching birds that weave intricate round nests.

The most elaborate woven nests are constructed by the tropical Afro-Asian weaver birds of the family Ploceidae. Some of the more primitive members of this family, which includes some 70 species, construct the usual cup-shaped nest but then add a roof —in the tropics the domed nest not only keeps out the torrential downpours but protects the bird from exposure to the high degree of ultraviolet light from the intense rays of the sun. Some species such as the baya weaver, *Ploceus philippinus*, have adapted vertical tubes, sometimes as long as 2 feet, through which the nest

Left: communal nest of the sociable weaver bird (*Philetairus socius*). Several pairs begin by building a joint nest attached to a stout tree branch. They build from the top downward, using twigs and the strong grasses of the dry South African savannas. The structure eventually houses some 20–30 couples in individual nests. It is extended year after year until some reach diameters of 15 feet. The flight holes, of which there can be as many as 125, are all on the underside. It is probably the most conspicuous birds' nest in existence, and its occupants face only one danger—that their fortress home may grow so heavy that the supporting branch breaks and the nest collapses.

is entered from below. This is an added protection against tree-dwelling snakes which would otherwise raid the eggs or young.

The habits of the common village weaver bird, *Ploceus cucullatus*, have been particularly well researched because it displays such great skill in nest building. The male first alights on a growing elephant grass plant or palm and bites a notch across the broad blade or frond. Then he tears the strip toward the tip of the leaf, takes to the air, and rips it free as he rises vertically. He chooses the tougher central strips to start the nest and wraps them around the fork of a twig to establish a firm foundation. Additional long strips are interwoven through a series of complicated acrobatic maneuvers by which the bird joins hanging strips into a ring, often working upside down. Finer grass strands are then added gradually using slipknots, overhand knots, loops, half hitches, and loop tucks. Eventually the spherical nest takes shape, and last to be woven is the tubular entrance. When his masterpiece is finished the male advertises for a mate by displaying himself and calling to females. When one accepts the nest she adds a lining of soft grass and feathers. As soon as his mate has laid the first clutch of eggs the male moves on to build a second nest for another prospective hen. He will continue, if conditions are favorable, to provide nests for several more.

Another bird that weaves or sews but is unrelated to the weavers is the tailorbird, a warbler of the genus *Orthotomus*, which is found in southern Asia. The long, straight bill of this common lowland bird is used as a needle to sew large leaves together, which then form a cup in which the actual nest will be built. The bill pierces the leaf and plant fibers and cotton scraps or cocoon silk are used as thread: each strand is drawn through the leaves and knotted to keep it from slipping out. If some

Above: the fairy tern or white noddy (*Gygis alba*) with its single egg. These seabirds often nest in the fork of a tree and frequently near other nests, as they like to breed close together. The chicks have very sharp claws, and very young chicks can hang upside down from a person's finger. The egg and chick are camouflaged by their color to protect them from predators.

threads are too short the bird twists several into one longer strand. Inside the completed leaf pouch the bird constructs a nest lined with wool, tree cotton, and other soft fibers. In this well-hidden nest four eggs are laid and the young nestlings can grow in comparative safety.

Other building skills abound in the avian world. Swallows and swifts, for example, build nests out of mud. The most unusual mud nest is that of some six species of Central and South American ovenbird of the family Furnariidae. Their nests look like primitive bakers' ovens. The building behavior of a breeding pair seems to be linked with the arrival of the rains and the formation of muddy clay. The oven is usually constructed on a tree branch or stump, but with the expansion of agriculture the birds have adapted to building on fence posts, telephone poles, rooftops, and even garden walls. Just as a house may be built of thousands of bricks, over 2000 small lumps of clay are collected by an ovenbird pair and molded by their beaks and feet to form the nest. In order to give greater cohesion and strength to the oven, plant material, straw, and cow dung may be mixed with the clay. The single curved corridor entrance leads from a small antechamber to a spacious brooding area. This chamber is lined with fine grass for the eggs. The structure when hardened provides good protection from enemies, but as soon as the chicks are sufficiently fledged, about 15 days after hatching, they leave the nest. This is because in the summer the nest becomes, ironically enough, too hot for the birds.

Another master builder in clay is not a bird but an insect, the British potter wasp Vespidae *Eumenes*. This solitary wasp molds a tiny, delicate pot of clay as a nest chamber for its eggs and larvae. They are attached, singly or in small groups, to a plant or hidden under the loose bark of a tree. If the clay is too dry the female moistens it with water regurgitated from her stomach. Her mandibles or "jaws" and front pair of legs are used to form pellets out of small bits of clay, which she then carries between head and thorax or breastplate to the building site. She rolls each pellet out into a strip with her jaws and legs and builds them up into a hollow flask shape. When she has formed the "neck" of the bottle she goes off hunting for caterpillar or beetle larvae which she paralyzes by stinging them. She carries each one back to her

Below: tropical mud dauber wasp (genus *Sceliphron*). It plasters a nest of mud on any suitable surface including the walls of houses, and sometimes even inside the house. Each species has its own particular prey (usually spiders or insects) which the wasp paralyzes and pushes into its nest.

Above right: the British potter wasp (*Eumenes*), architect of a beautiful round clay nest in the shape of a flask. Each has a short neck and flared rim. The wasps' nests can be found on heather growing on dry sandy heathland.

Master Builders in Mud and Clay

Left: worker honey bees (genus *Apis*) filling wax storage cells with nectar gathered by the bees from flowers. The hexagonal cells of the honeycomb are a marvel of purpose-built construction.

pot and pushes it in through the neck. After filling the pot with one large or several small caterpillar bodies, she inserts the tip of her abdomen down the neck of the pot and lays a single egg, which is covered with a sticky liquid. This quickly hardens into a thread by which the egg is suspended from the upper wall of the pot. A last clay pellet is then used to seal the pot after which the female goes on to build others. The hatched larval potter wasp has sufficient food to see it through to maturity, at which point it breaks out of its cell.

The smith or blacksmith frog, *Hyla faber* of Argentina and Brazil, builds a clay wall for its nest. The name of this tree frog refers to its call, which sounds like a blacksmith hammering on metal. At breeding time the male builds a circular wall of clay near calm shallow water, enclosing an area about a foot in diameter. He uses his hands to pick up lumps of clay which he smooths and molds before adding them to the wall. After about two days' work the male sits inside his "castle wall" and croaks to attract a female. She hops over the wall, mating and spawning take place, and then the adults depart, leaving the tadpoles to hatch out in a well-protected nursery.

Honey bees of the genus *Apis* have adapted to life in hives provided by beekeepers, but the wild species make their nests in hollow trees or rock crevices. The bees' honeycombs are made of wax secreted from between the abdominal plates of worker bees, which are infertile females. These storage areas are composed of individual cells separated by vertical walls which form hexagonal prisms, each of a regular size and shape. The worker bees insure that stored honey or pollen do not spill out by tilting the cells slightly, and when full they are sealed.

For centuries scientists, architects, and civil engineers wondered why honeybees had evolved the innate programing which tells them to build only hexagonal storage cells. By a process of trial-and-error it was discovered that circles would result in areas of unusable space between the individual structures, a situation which any good architect tries to avoid. Three-, four-, five-, and six-sided cells, however, contain equal space. Why then was the

Complex Homes of Ant Species

six-sided hexagonal cell chosen? The solution to the puzzle proved to be that the circumference of a hexagon is the shortest in length, and if a bee is making hundreds of cells in its life, the cell providing maximum area for the least investment of building material would be the best one to build.

Bees have evolved a building technique of astounding precision in that all the cells are of a regular thickness and are made in a size which fits their purpose. Those intended to house male drones are slightly larger than those for female workers, but none are larger than they need be.

Another mystery that puzzled scientists for many years was how the bees all worked together so quickly and efficiently without needing a supervisor to direct and coordinate their efforts. Eventually experiments showed that the bees orient the vertical line of the comb with the natural north-south line of the earth's magnetic field. This means that even if hundreds are working together their basic blueprint is instinctual. The result is that they can build a well-ordered structure very quickly. What remains a mystery is how the bees are able to perceive the magnetic field of the earth.

All of the 6000 or so species of ants are social insects, meaning that they live in organized colonies, but the more primitive species merely dig passages and cavities in the soil. The most imposing anthills—large mounds of earth, conifer needles, dry stalks, and broken twigs covering underground nests—are constructed by European wood ants, *Formica rufa*. Smaller, less conspicuous anthills are built by other species, but although ants are related to bees their homes have nothing in common with the bees' waxen structures. The mounds are worked by hundreds of thousands of ants, whose excavations may reach as deep below the ground as the visible part of the anthill rises above the ground. To add strength to the walls below ground level the ants mix a kind of mortar from soil and body secretions. A new nest is often begun inside a rotting tree stump, and the workers hollow out the first passages and chambers in the decaying wood. In a short time the stump will be completely covered by fir needles and small light twigs.

Below: weaver ants (*Oecophylla smaragdina*) building a nest, which consists of living, undetached leaves held together by a dense silky web. To begin they attach themselves firmly to a leaf with the sharpened claws of their six legs. Then they seize the edge of another leaf with their mandibles and try to pull it closer by moving their legs, one after the other, further back.

Below right: when the distance between the two leaf edges is too great, other ants join in to form a living chain to bridge the gap.

Although a completed mound may look static, it is actually being moved continually. In one experiment a quick-drying blue dye was sprayed over a mound, a type which would not make the twigs and pine needles stick together. In only four days the mound was back to its usual brown color, and the blue material was found about 4 inches below the surface. Different color dyes were sprayed in sequence, and the same thing happened, but after a month the dyes began to reappear in the order of their original application. The ants must instinctively know that if the material of the nest is not continually turned over and aired the humid conditions inside will encourage mold.

Ants that live in temperate areas require as much solar heat as possible to keep their nest heated, and a mound catches far more of the sun's rays than would a nest at ground level. Another form of heating is provided by large numbers of ants sitting exposed to the sun—when hot they return to give off this heat inside the nest while another group takes over the chore of sunbathing.

The most expert civil engineers among social insects are the termites of the order Isoptera, of which there are some 1700 species. Most of these live in tropical and subtropical regions, but a few are also found in southern Europe and North America. A termite home may contain over 10 million individuals, far surpassing the largest ant nest in size, and in some tropical regions they may soar to over 20 feet in height. When large numbers of such nests are found close together the landscape can look like a city in a science fiction novel.

The exterior of a termite mound, made from a cement of termite feces, is as strong as reinforced concrete and prevents the temperature inside from getting too high. The hard shell also keeps out some predators, although animals such as anteaters, aardvarks, and pangolins can break through with their claws. A termite mound shows far greater planning and is more of an architectural structure than an ant's nest. The focal point of a termitarium is usually a royal chamber where the queen and king live surrounded by numerous other chambers and galleries. At one time it was believed that the queen, who grows to an enormous size compared to the workers, was imprisoned in her

Below left: the next step. In order to join the two leaf edges, each worker ant holds a larva between its mandibles. Using the larva as a weaving "shuttle," the workers press the larvae's mouths against the leaf surface first on one side and then on the other, squeezing them with their mandibles. This makes the larvae discharge some of their glandular secretion, thereby stitching the two leaf edges together with a multitude of silken threads.

Below: the finished nest. The use by ants of their own larvae as both spindle and shuttle is one of the most remarkable instances of the use of tools by animals.

chamber for life because she is much bigger than the openings which lead into the chamber. However, if danger threatens the royal cell she can with great effort squeeze into another part of the nest. If a termitarium is broken open, the spongelike interior can be seen. The atmosphere is permanently moist, warm, and stale with a high percentage of carbon dioxide, 5 to 15 percent. A man confined in such an atmosphere would soon lose consciousness. Some parts of the nest are designed as storage chambers for leaf sections, and some termite species even build food-growing chambers where they cultivate fungi on leaf mold.

Termitaria are designed to suit the conditions of the environment. For example, certain species of the genus *Cubitermes* roof their homes as a protection against tropical downpours. They look like miniature pagodas. A particularly remarkable engineering feat is achieved by the compass termites, *Amitermes meridionalis*, of Australia's arid and treeless northern steppes. Their towering nests may be 10 to 12 feet high and up to 10 feet long, shaped like a flat wedge. The amazing feature is that every compass termite structure is aligned so that the two short sides face due north and south. Thus the surface exposed to the scorching heat of the midday sun is small. The long sides catch the cooler rays of morning and evening. How these insects achieve their architectural orientation has not yet been determined, but laboratory tests have shown them to be sensitive to the earth's magnetic forces. Termites also incorporate an air ventilation system into their homes so that stale air is slowly removed and replaced by small quantities of fresh air.

Nest-building is not as important to mammals as it is to birds or other animals. This is not only because the mammalian mother is warm-blooded, as are birds, but is also due to the strong bond between mother and offspring—she is particularly attentive and suckles, warms, and protects her babies. However, although nest-building is not as widespread in mammals as in other animal groups, quite a number are builders.

The most beautifully constructed mammalian nests above

Astonishing Termitaria

Above: larvae of the leafcutter bee (*Megachile centuncularis*), a solitary bee that cuts out neat pieces from leaves with which to line the cells of its nesting burrows. The result is a series of larval cells stocked with food and looking like small cigar stubs.

Opposite: large termite mound, the result of far more architectural planning than an ant's nest. Different mound-building species have different techniques, and not all are equally well organized. Termites living in arid zones do not build overhanging, mushroom-shaped roofs on their mounds.

Below left and right: the larval casings of two different caddisfly species. When disturbed the larvae withdraw their heads and legs inside and hold on with two small hooks. Their soft bodies would otherwise end up in a predator's stomach.

ground are built by the Old World harvest mouse, *Micromys minutus*, and the hazel dormouse, *Muscardinus avellanarius*, of Europe and Asia. Both rodents, neither more than 6 inches long, weave nests out of blades of grass in fields, hedgerows, or thickets. The harvest mouse, smallest mammal in Britain, uses its front paws and incisor teeth to construct a tiny round nest 3 inches in diameter and 1.5 to 3 feet above the ground. It usually incorporates stiff material such as oat stalks for support. Gripping with hind legs and one front paw, it can also use its tail as an extra "hand." The harvest mouse can finish a nest in five to ten hours, after which it becomes the nursery for a litter of five to nine babies. The youngsters grow and develop rapidly and after about two weeks they are ready to leave the nest, by which time they have probably destroyed the compact structure. The mother must construct a new nest for each litter, and several may be required during the breeding season from April to September. Male and young harvest mice build themselves simpler, less dense nests for sleeping.

Above: dainty nest of the European or Old World harvest mouse (*Micromys minutus*), which can stand comparison with some of the most skillfully woven birds' nests. Its habitats are fields of cereal plants, meadows with long grass, and stands of sedge and reeds on the shores of lakes, marshland, or grassy scrub. While building its nest the mouse grasps thin leaves of oats or blades of grass or sedge and pulls them through its mouth, reducing them to long thin shreds with the tips of its incisors. Once a base has been formed by interlacing the leaves of stalks standing together, the rest of the nest is built up in a similar manner and finished off with a domed roof. The completed nest is a firm dense ball with a lateral entrance hole, and the inside is padded with seed down, panicles of flowers, and shredded leaves.

American pack rats of the genus *Neotoma* build large, rather coarse nests from twigs and brushwood which are more like small lodges than nests. Sometimes as tall as 3 feet high, they contain a living chamber, nursery, storage chambers, and even a lavatory. The bushy-tailed wood rat, *N. cinera*, even adorns the outside of his home with shiny objects such as tin cans, spoons, broken glass, and other rubbish from nearby farmyards or camp sites. Why pack rats have this curious drive to collect is not known, but they always leave something in exchange.

Tree squirrel nests are similar to those of harvest mice and dormice but are much less well organized. Species include the gray squirrel, *Sciurus carolinensis*, of eastern North America and the Eurasian red squirrel, *S. vulgaris*, and their nests are called *dreys*. They usually measure some 20 inches across and a foot or so high. Several are built by one squirrel in adjacent trees, and each serves a different purpose, one being used as a nursery. Although twigs and branches are the usual building materials, sometimes the discarded nest of a crow or nuthatch may be patched up for reuse. The drey is lined with grass, thin strips of bark, leaves, and moss, and the floor is padded with more shredded grass, lichens, and moss. A small entrance hole is left on one side. At the beginning of the breeding season a male and female may share a home but as the end of the gestation period approaches the pregnant female chases off her mate. She cares for her young alone and is an efficient mother, keeping them clean and fed as well as cleaning the nest daily.

Probably the best known of mammalian builders is the beaver of which there are two species: the Canadian beaver, *Castor canadensis*, and the European beaver, *C. fiber*. Second in size among rodents only to the capybara of South America, this creature was once distributed in hundreds of millions over North America from northern Mexico to the tree line of Canada as well as over Eurasia from Lapland to Italy and Britain to the eastern Soviet Union. Today it is found only in small and isolated areas due to its continuous slaughter from the 10th to the 19th century. It was coveted both for its beautiful fur and also for a glandular secretion, known as castoreum, which was used as a cure-all tonic. Beavers eat the bark and sapwood of small aspen and willow branches, and their occasional forays into cultivated forests have earned them some enemies.

The beaver's adaptations for its life of building are numerous. The large front incisor teeth grow continually throughout its life, renewing the enamel worn away by chewing and gnawing wood. The beaver's hands are also indispensable, although the thumb is poorly developed. Its little finger can be used like man's opposable thumb, and the hands are used to pick up sticks and to shovel up mud or debris from the beds of ponds. Beavers can also carry large piles of twigs between chin and arms, use their arms to lift loads, or walk upright on their hind legs.

Beavers live in family groups made up of a monogamous pair and their young, which stay with their parents until they are about two years old and are sexually mature. Their home is usually a lodge built in a "beaver pond," which is constructed by damming a stream or slow-moving river until it overflows, but if

An Instinct to Construct?

Below: European beaver (*Castor fiber*) outside its lodge. Many animals can construct homes from branches and twigs as beavers do, but using these materials to build a dam across a river or stream, that transforms a shallow stretch of water on the upstream side into a calm lake of sufficient depth for their lodges and their swimming and diving activities, is an achievement matched by no other animal. The details of the design are flexible, as is the size. Large dams are maintained by generations of beaver families over a period of many years, perhaps even centuries.

What Triggers These Activities?

Below: skeletons of Radiolaria, unicellular floating plankton of warm seas, which extract silica from seawater. From this substance they fashion internal bodily. structures of marvelous beauty which serve as protection and support. In most species these look like little helmets or intricately interlaced, latticed balls. The Radiolaria group dates back to the Precambrian era about 700 million years ago, and the beds of many tropical seas consist predominantly of their skeletons.

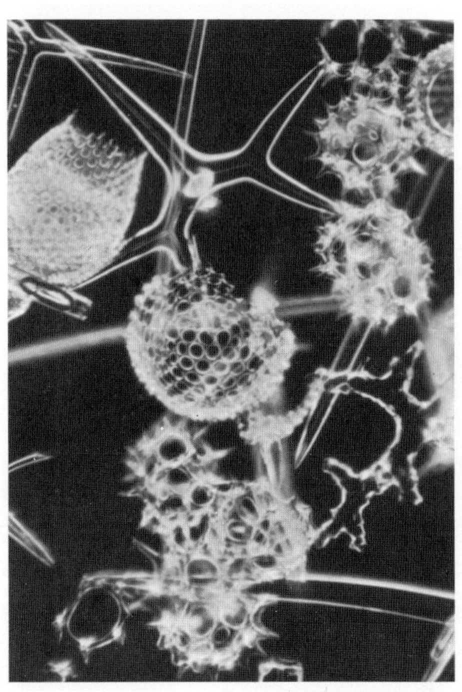

the river is wide and the water level constant they may live in a burrow in the bank.

To construct a dam in shallow water beavers use a technique unique to them. First strong sticks are rammed into the stream or river bed and twigs are pushed into the gaps. Forked branches are used as support to prevent the edifice from being swept away, and the heaviest branches are placed on top to weigh down the others. Boulders, growing trees, or any other materials near to hand may be used, attached by long branches or thin trunks, to anchor the dam. Heavy stones are dragged to the site if reinforcement is required, and gaps in the dam are filled with reeds, tiny twigs, mud, and clay. Dam size varies depending on the width of the stream or river to be blocked—the largest on record in the Soviet Union was some 400 feet long, 3 feet high and 2 to 3 feet wide. Probably the biggest in the world was a dam built across the Jefferson River in the state of Montana, which was over 2000 feet long and could bear the weight of a man traveling across it on horseback.

Beavers are also skilled hydroengineers, as they alter their dams according to the water level. If this rises after a torrential rainfall and the lodge is in danger of being submerged, the beavers will lower the crown of the dam so that more water flows over it and the level of the beaver pond falls. Breaches in the dam are also quickly repaired.

Preferred trees are the softwood poplars, aspen, and willows, although the beavers' excellent chisel-shaped incisors can tackle hardwood oaks as well. A thin branch can be bitten through in minutes, whereas a stouter trunk or branch is tackled from all sides—often two beavers will take turns doing the chewing. They seem to know instinctively when to run for safety as the tree topples, although some, probably among the younger and less experienced, meet with accidents. Beavers may also construct canals to transport felled logs and branches to their pond if the trees are not close to the pond.

The beaver's home is the lodge where the kits are born and the family shelters through the severe part of winter. It looks like a straw beehive in shape, half submerged in the pool, and is constructed of interwoven branches daubed with mud or clay. No mud covers the center of the dome, which acts as an airshaft. Inside is a large, well-insulated living chamber reached by underwater entrance holes, which are sited low to keep them from freezing over. Here the family, sometimes three generations, spends the winter, surviving on stocks of wood piled into any available space inside the lodge or placed close to an entrance. In spring the mother evicts the rest of the family and prepares a bed of woodshavings on which the young kits are born.

Although it may seem that beavers must be very intelligent to possess such skills, this is not actually the case. Though their brains are quite large, their craft is a result of inherited behavior patterns. Young beavers born in captivity will fell trees and erect lodges if materials are at hand. In their natural habitat, however, they no doubt perfect their building skills by experience and by watching their elders.

Surprisingly enough the apes, man's nearest relatives, do not build complex or permanent homes in spite of their high degree

Left: chimpanzee (*Pan troglodytes*) in its sleeping nest. The ape first chooses a firm foundation in the crown of a tree and then bends over small branches onto this foundation, keeping each one in place with its feet. Finally the little leafy twigs growing around the rim of the nest are tucked in and the chimp lies down, pressing down the whole cushion with its body. Building sleeping nests is instinctual behavior, as shown in the case of an adult orang-utan in captivity in the Munich zoo. He went through the motions of building a nest by picking up nonexistent "twigs," bending them toward him, and carefully pressing them down with the back of his hand.

of intelligence. The chimpanzee, *Pan troglodytes*, will in its native habitat (the rain forests of West and Central Africa) use a large leaf as a sunshade. It also weaves itself a sleeping nest, like a crude hammock, from living branches at the top of a tree, but this is newly built each evening at dusk in about five minutes. The British zoologist Jane Goodall observed chimpanzees in the wild near Lake Tanganyika in Central Africa over a period of several years. She noticed that often on settling down for the night a chimp would reach out to pick a handful of leafy twigs, which it used for a headrest. She also observed that healthy chimps never fouled their nests, always relieving themselves over the side.

A female with young will take more time to build a large nest that can accommodate herself and her baby. It sleeps with her for about four years, and from the age of about 10 months the youngster practices nest-building skills; sometimes it bounces on a nest using it as a trampoline. By the time the youngster is four or five years old it has acquired the skill of constructing a nest, having practiced so often. Orang-utans and gorillas also make fresh nests daily, but the larger gorillas often build them on the ground as they are too heavy to sleep in trees.

All forms of home-building, whether the large complicated structures of termites or the tiny, round nests of harvest mice, are the result of innate behavior patterns specific to each kind of animal. These have been adapted by evolution in the same way as have the animals' other behavior and anatomy. Each pattern is triggered by specific factors that, in this case, urge an animal to begin searching for a suitable building site and the necessary materials. Most of these factors are still unknown as few analytic studies have been made.

Birds are probably the best understood group so far: the reproductive cycle of the female canary, for example, is affected by the sun, the continued presence of a male, and internal hormonal changes. Each step, including nest-building, is triggered off at the appropriate time and in the proper order. But only after much more patient research will we begin to understand the intricate mechanisms that send animal architects into action.

Chapter 11
Domestication of Animals

In the beginning there was the family of man, but as civilization evolved the group expanded to include dogs, cats, and a veritable zoo of domesticated animals—pigs, sheep, oxen, horses, cattle, goats, and even reindeer, camels, and gazelles. What led men to trap and tame wild animals for their own convenience, and eventually to breed them for particular characteristics? What exactly is the special relationship between man and his domesticated companions? This chapter also explores how domestication affects animals and what makes some species more amenable to being tamed than others. But perhaps the most interesting question, and the most difficult to answer, is how both man and animal have been changed by their experience.

It is impossible to say exactly when meat-eating became important in the life of man, but modern thinking holds that about 5 million years ago hunting techniques and skills were gradually developed. It was probably not until 2 million years ago that big-game hunting took place on a large scale. Modern man, *Homo sapiens sapiens*, may have emerged around 50,000 years ago, but for no period has evidence been found that man had any kind of close relationship with wild animals other than hunting and killing them. The first wild animals may have been tamed by Stone Age men about 10,000 years ago when men began to live in permanent rather than nomadic settlements. Gathering wild the natural products of their surroundings became less important as they developed primitive agriculture and animal husbandry. Stone Age men then learned to tame wild animals in order to benefit themselves.

There are two groups of domestic animals: those that submitted to a more or less voluntary dependence upon man, and those that were enslaved and then tamed by him. Dogs and cats belong to the first group, while various hoofed mammals belong to the latter. This second group includes farm animals such as cattle, horses, pigs, goats, and poultry as well as, in particular regions, reindeer, camels, and certain antelope species.

The first animal to be domesticated was probably the dog, *Canis familiaris*, whose remains have been found with the bones of Mesolithic (middle Stone Age) man. The domestic dog is a

Opposite: farm horse at work in the Ardennes region of France. Domestication of the horse has made a tremendous impression on the history of mankind. Horses may have been used to pull chariots before they were ridden by individuals. In time, however, they were employed as pack animals for heavier transport, especially for cavalry. The development of the modern horse is a matter for debate, though it seems likely that several wild subspecies were distributed across Europe and Asia and that the earliest domesticated breeds reflected the differences between them.

The First Dogs

canine, related to wild animals such as wolves, jackals, foxes, coyotes, and hunting dogs. They all possess a recognizably similar body shape; in behavior they vary between a majority which live and hunt in packs and the solitary minority of foxes and coyotes. Which wild canine provided the ancestral stock for the breeds of dog we recognize today?

It was once thought that both the jackal and the wolf were part of the ancestry of the domestic dog, but today the role of the jackal is considered doubtful. The wolf, *Canis lupus*, is the only species to show both the intelligence and social behavior characteristic of the domestic dog.

The first dogs were probably descendants of one of the smaller southern races of wolf found in India and formerly in China. One trait which distinguishes dogs from their wild relatives and forebears is the upturned tail, and in the Stone Age village of Jarmo in Iraq figurines were found illustrated with dogs with curled tails. The oldest artifacts date from 6500 B.C. Later, the bones of a domestic dog associated with a British Stone Age culture were found in Star Carr, Yorkshire, dating from 7500 B.C. German finds of about the same age and discoveries in Denmark that date from at least 1000 years later have also come to light. This evidence seemed to imply that the domestication of the dog must have occurred after the end of the last Ice Age in about 8000 B.C.

However, within the last decade Dr. Barbara Lawrence of Harvard University in Massachusetts has identified fossil material excavated from Jaguar Cave, a Stone Age North

Below: reconstruction of a Neolithic farming settlement by the Czech artist Zdenek Burian. The huts are rectangular and spacious. The corn crop is cut with sickles, and the grain is ground on stone slabs. Livestock have been domesticated, and the art of making clay vessels has become well established.

Left: Assyrian relief from Nineveh showing the king's hounds being walked in the royal park. Nineveh was the splendid capital of the Assyrian empire when it was at its peak in the 7th century B.C.

American Indian site in Idaho, as domestic dog bones dating back to 8300 B.C., almost a thousand years before the Star Carr dog. In fact, not one but two races of dog could be distinguished, one medium-sized and one quite large. The ancestors of these Indians had migrated from Siberia a few thousand years earlier across a land bridge over the Bering Strait which no longer exists. They may have brought the dog with them from the Old World, which would imply that the dog had been domesticated much earlier. More recently dog bones have been found in Czechoslovakia which come from late Paleolithic archaeological layers, and they seem to date well back into the Pleistocene epoch, probably to about 10,000 B.C.

The wolf has been considered the prime candidate for ancestor of *Canis familiaris* for several reasons. Both dog and wolf show great individual and geographical variation. The wolf's social structure also makes it an ideal animal for domestication. Wolves in the wild live in groups, and within the unit most are subordinate to the dominant leader of the pack. When domesticated they accept man as a surrogate leader and submit to him without difficulty.

The first dogs to be domesticated may have been tiny puppies whose mother had been killed by a hunting party. Modern research has shown that mammals born with their eyes shut, such as dogs and wolves, imprint or fix as a model the image of the first moving thing they see upon opening their eyes. Young puppies would thus become imprinted with the image of a man and would identify themselves with his species. Wild wolves can

Above: the wolf (*Canis lupus*), considered the prime candidate for ancestor of the dog. Once widespread over Europe, most of Asia, and North America, the wolf's range was probably greater than that of any other land animal. Today there are two distinct species of wolf—the gray or timber wolf of northern Europe, Asia, and North America and the red wolf, which is restricted to the south-central United States. However, there are numerous local races.

even become part of a human "pack" if the puppies are taken at an early enough age.

The domestic cat, *Felis catus*, also has many relatives in the wild, but the European wild cat, *F. sylvestris*, and the bush cat of Africa, *F. lybica*, are considered its direct wild ancestors. The domestic cat has the skeletal structure of its wild cousins but greater variety of coat texture, color, and pattern. The first cats to be domesticated may have been kept to fight the small rodent pests that attacked early man's stores of grain, and even today the cat, a loner in the wild, rarely becomes as attached to one person as would a dog. A cat usually identifies more with the home where it is fed rather than with the people that live in the house. In early Egyptian times, before 1600 B.C., cats were wor-

Above: bronze Egyptian statuette of a cat from the Saite period, around 600 B.C. The cat is wearing earrings made of gold. This may have been a sacred idol for worship or a representation of a domesticated animal.

Right: Egyptian wall painting from the tomb of Nebamun at Thebes. This painting depicts the inspection of domesticated cattle and dates from around 1400 B.C.

Solitary Creatures

shiped and mummified—the word "puss" probably comes from the Egyptian goddess called Bast or Pasht. To harm or kill a cat was punishable by death.

Although domestic cats have lost the savagery of their origins they remain aloof and independent, solitary creatures within a group environment. Cats cannot be trained to perform specific tasks because they are not wholly dependent on man—most are quite capable of capturing birds or small mammals if a meal is not forthcoming. *Feral* cats, which have "gone wild," are a feature of most towns and have become a problem in many areas.

The domestication of most other animals probably accompanied the development of agriculture. Wild cattle and buffalo may have raided crops and in the process of chasing them off men

The Last of the Aurochs

may have captured some. The earliest evidence of domestic cattle has been found in Greece and Turkey, dating back to 6000 B.C. Cattle bones have been found at Suberde, a Stone Age village at the foot of the Taurus mountains in southern Turkey, from which a picture can be built up of what people ate. Between 7000 and 6000 B.C. the proportion of beef in man's diet rose from 14 to 30 percent while goat and mutton fell from 70 to less than 50 percent. This change suggests that a wild species of cattle had been domesticated during this period while goats and sheep, which were still wild, were no longer needed for food to the same extent.

However, one lingering mystery was why, for each complete leg structure, several feet were found. An explanation may be that the Stone Age Turks had difficulties similar to those of the North American Plains Indians, who hunted bison and often killed the enormous beasts many miles from their homes. To haul the complete carcass back would have proved difficult and arduous, so the Indians skinned the animal on the spot and stripped the meat off the long leg bones, wrapping it up in the skin to drag it to their camp. They always left the feet attached to the skin to use as handles for pulling the load along.

The wild cattle that the people of Suberde hunted were aurochs, *Bos primigenius*, a now-extinct species of European wild ox. These bovines were deep-chested, agile, and aggressive. From medieval descriptions and cave paintings we know that the bulls were rich black with a yellow stripe down the spine, while the cow was smaller, with a red or dun-colored body, darker head, neck, and legs, and a dark stripe down the spine. The last auroch died comparatively recently in 1627 in a Polish park. They once ranged from northwestern Europe through the Mediterranean to the Near East and well into Asia.

Above: reconstruction of the auroch (*Bos primigenius*), an extinct species of European wild ox. The last known specimen was kept in a zoo in Poland, where it died in 1627.

Nearly all the auroch bones found at Suberde came from adult bulls. A possible explanation relates to the herding behavior of wild cattle which may have been similar to that of red deer today. For much of the year red deer stags live in herds separate from the hinds (females) and their young. Females are very wary and timid, retreating quickly from danger to try to protect their young. Stag herds tend to face intruders boldly and are prepared to use their antlers in defense of their territory. If bull aurochs had a similar behavior pattern, the bulls would be easier prey for hunters than cows and their young.

About 100 miles from Suberde is Catalhuyuk, the largest Neolithic site yet known in the Middle East. Excavations by the British archaeologist James Mellaart from 1961–5 showed that, although the date is the same as Suberde's, here the aurochs were tamed. All the bones found are in correct proportion—every leg has one foot—which suggests that the animals were slaughtered on the site. Many of the remains are of females and young animals. Even the bulls were smaller by about 15 percent than the local wild cattle. Decrease in size is often a result of domestication in certain herbivores, because most animals in captivity have a smaller feeding range. If not enough grass is available the herbivore, especially a larger one needing greater amounts of food, would slowly starve. Under these conditions natural selection would operate in favor of smaller animals.

The earliest evidence of domesticated cattle found in Europe dates from the early Neolithic or polished Stone Age period. Probably a long-domesticated strain introduced from Asia, this slightly built breed persisted in Britain until Roman times and is sometimes known as the Celtic shorthorn.

Cattle shapes altered as man chose animals for breeding which showed certain desirable characteristics. The Romans developed breeds of cattle that gave good beef, which resulted in shorter, stockier limbs and meatier hindquarters, a distribution of meat which reversed that of the aurochs. In Britain the wild herds of White Park cattle, of which one herd still exists at Chillingham Castle in Northumberland, are descendants of Roman cattle that became feral when the Romans left Britain.

There are two quite distinct types of domestic cattle, the European and the zebu, *Bos indicus*, of southern Asia. Its fat hump easily distinguishes zebus from European cattle, and this was probably encouraged in breeding for no reason except that it was regarded by herdsmen as decorative. Over several thousand years the zebu was selectively bred for the hot, desert conditions of China and Pakistan while the European breeds departed less from their ancestral aurochlike features. The zebu sweats less and requires less than half the amount of water as European cattle of the same size do in order to survive. European cattle have more fat distributed evenly beneath their skin in addition to the "marbling" within their muscles, thus improving their heat insulation —important in the cold winters of temperate areas. By contrast the zebu has most of its fat, apart from the hump, around its internal organs.

Below: Chartley bull, one of a herd that has been kept and bred in captivity since 1248. They are thought to be the descendants of the original white cattle introduced into Britain by the Romans, possibly as sacrificial animals for slaughter in their religious ceremonies.

When Did Sheep Evolve?

Right: zebu (*Bos indicus*) or Brahman cattle. Originating in southern Asia, possibly in India, their wild ancestor is unknown. The zebu's prominent shoulder hump is an enlarged muscle rather than solely a store of fat as is sometimes claimed. Its horns are more upright than those of the aurochs. Some experts claim it was domesticated from the wild gaur or banteng of southeast Asia, but it differs from these in its long slender face and other features. Domestication may have occurred earlier than for Western breeds of cattle.

It was because of these features that zebu breeds were introduced in America and Australasia a century ago. In recent decades, however, specialized breeding programs in many parts of the world have developed new breeds for particular environments, crossing breeds of European cattle with zebus to get the hybrid most suited to local conditions. In Europe cattle are bred for four main purposes—meat, milk, labor, or a combination of the first two—but in some countries the hides, fat, bones, horns, and even the dung are used, the latter as fuel or for building primitive huts.

Although today cattle are the most important group of domestic animals, goats and sheep were also tamed very early in man's history as an agriculturalist. Evidence of bones and horns indicates that domestic goats were present around Jericho in the valley of the lower Jordan river as early as 7000 B.C. At this time the wild bezoar goat, *Capra aegagrus*, was common in the mountainous area to the east, so the goats found in the flat grasslands of the Jordan valley were probably brought from elsewhere by herdsmen, and goats may have been domesticated as early as 10,000 B.C.

Grasses by this date had been cultivated for several thousand years, and it is quite possible that the resourceful goat was the first animal to invade man's crops, finding easy pickings in the dense stands man grew for his own food. Perhaps at first several adults were slain and this led to orphaned kids. These would have been taken back to the camp to be fattened until they were large enough to be worth killing. As with the offspring of many herbivores, whether kid, lamb, calf, or fawn, if reared by man from a young age they soon become tame and attached to their keeper. It would quickly have become obvious that keeping a tame herd of goats would be better than traveling long distances to hunt them.

Evidence that wild goats were also hunted comes from Paleolithic cave drawings. No wild sheep are found in the paintings, though, and only about 25 sheep bones have been found that

Left: Barbary sheep (*Ammotragus lervia*) of northwestern Africa. Also called the aoudad from the Berber word, they live on the craggy peaks of the Atlas Mountains. Their natural predator is the Barbary leopard, which is becoming increasingly rare because of the demand for its pelt.

date back before 10,000 B.C. Even these bones might actually belong to goats. Because animals that were hunted, such as aurochs and wild goats, generally appeared in cave paintings, it has been suggested that wild sheep did not yet exist in Europe or the Near East during the Paleolithic period. They may not have evolved until between 15,000 and 10,000 B.C.

From chromosome evidence it seems possible that sheep may have arisen from goats through genetic mutation. The wild goat has 30 pairs of short chromosomes, while of the domestic sheep's 27 pairs, 24 pairs are short and 3 are long. Two short pairs joined together match a long pair in size and shape. Perhaps the sheep's three pairs of long chromosomes are the result of short pairs becoming attached at some time in the history of the wild goat. Certain existing species of wild sheep have intermediate numbers of chromosome pairs: Barbary sheep (*Ammotragus lervia*) and urials (*Ovis orientalis*) have 29 pairs out of which only one chromosome pair is long, and the argali sheep (*O. ammon*) has 28 pairs, two of which are long.

From the chromosome evidence the earliest wild sheep were probably the urial, which ranged from Mesopotamia to what is now Bangladesh, and the argali or Marco Polo sheep which lived further east and reached what is now the Soviet Union. Closely related wild species such as the Dall (*O. dalli*) and bighorn (*O. canadensis*) have reached North America. The most recent species of wild sheep, the mouflon (*O. musimon*), has been suggested as the immediate ancestor of most domestic breeds of sheep since its chromosome configuration is the same, but its range has been seriously depleted in recent times, and it is now found wild only in Corsica and Sardinia in the Mediterranean west of Italy.

There is one main drawback to this theory. The range of the Barbary sheep (with its 29 chromosome pairs, of which only one is long) is the most westerly of early wild sheep—it is found along the northwestern coast of Africa—but it is not closely related to the mouflon (which has 27 chromosome pairs of which 3 are

Above: urial (*Ovis orientalis*). None of the wild sheep has a wooly coat comparable to that of domestic sheep. Instead they have coats of coarse hair ranging in color from creamy white to brown. The coat of the wild sheep is therefore not unlike that of a goat. However, sheep have characteristically narrow noses with concave foreheads, as opposed to the convex foreheads of goats.

long). The problem, therefore, is to explain how, if sheep evolved as recently as 15,000 to 10,000 B.C., they could have expanded their range to include Barbary and North America in such a short time.

The Enigma of Cross Breeding

The answer may lie in the fact that sheep are better suited to a herded life than goats and may have been quite rapidly domesticated. The wild breeds of sheep that exist today may be simply descendants of some of the earliest tamed sheep which escaped, returned to a feral state, and evolved differently in the wild. Sheep prefer flat pasture to rocky ground and feed on grass rather than bushes. They herd more readily and are much less nimble than fleet-footed goats. The wild urials of western Asia would have been easily herded, and migrating nomadic tribes could have helped to disperse the sheep from there. Although little archaeological evidence has yet been discovered, it is possible that herdsmen took some sheep with them across the land bridge that existed until about 7000 B.C. between Siberia and Alaska, and that feral descendants of these urials became the bighorns of western North America.

The fact that it is very difficult to trace modern breeds back to the supposedly original three wild species—the urial, argali, and the more recent mouflon—lends weight to this theory of the sheep's origins. Different types of sheep can breed quite readily in spite of the minor differences in their chromosome numbers. However, sheep and goats cannot be bred, nor can domesticated sheep be bred with Barbary sheep. But it is possible that at one time these animals could cross breed, and if so it would also explain why the identification of bones from this period is so difficult.

The early stages in the domestication of the horse, as of many other animals, are largely conjectural. Man tamed the horse between 3000 and 2000 B.C., and it is possible that three separate species were domesticated at about the same time. In North Africa the wild ass was tamed for use as a beast of burden by the

Opposite: mouflon (*Ovis musimon*), the only European wild sheep. It stands about 2.5 feet high at the shoulder and is reddish brown with a whitish saddle patch and black markings on the limbs. The horns are usually found on males only and form a close spiral with the tips sometimes curving slightly inward as here. Occasionally a female may have short, slightly curved horns. The muzzle is narrow, the ears are pointed, and there is a fringe of long hair down the front of the neck. The mouflon is found truly wild only in Corsica and Sardinia but has been introduced into Germany, France, Switzerland, Austria, Hungary, and Italy.

Left: domestic sheep. Sheep crop short plants and grasses by biting with the teeth in their lower jaw against the gums of the top jaw, which has no teeth at the front.

Ancestors of the Horse

Right: Przewalski's horse (*Equus caballus przewalskii*), probably the ancestor of today's domesticated horses. Named after Nikolai Przewalski, a Russian traveler and explorer who discovered them in the late 1870s, they are the only surviving wild horses. The few that remain in the wild, in Mongolia, are now protected by law. They are smaller than domestic horses and like zebras have an erect mane.

people of the Nile valley; in Sumeria to the east the Asiatic wild ass was trained to pull war chariots which used the newly invented spoked wheels. However, true domestic horses probably developed from a species of wild horse that has all but disappeared in the wild—Przewalski's horse, *Equus caballus przewalskii.* Although it survives in zoos in herds of over 250, only a small number survive in the wild in a few isolated corners of Mongolia. Because today there are three identifiably distinct groups of horse breeds—exemplified by the light Celtic pony of Great Britain and Iceland, the delicate and graceful Arab, and the heavier Western shire and Belgian horses—it is not universally agreed that Przewalski's horse is their sole ancestor.

The first horses to be domesticated were used in warfare by the nomadic tribes of the Asian steppes because they were stronger and faster than domesticated asses. However, not until the Chinese invented the stirrup in about the 5th century A.D. were huge cavalries formed, after which battle horses were selec-

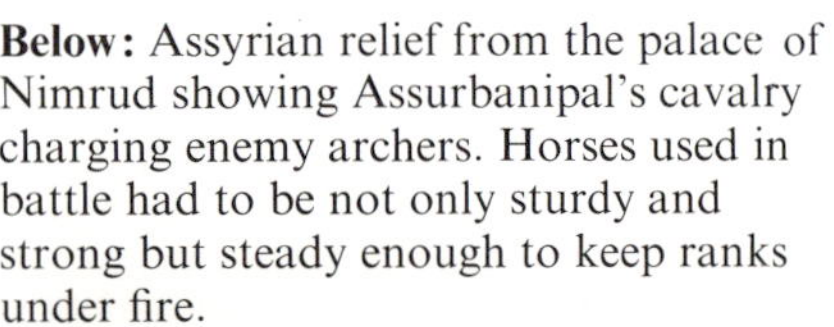

Below: Assyrian relief from the palace of Nimrud showing Assurbanipal's cavalry charging enemy archers. Horses used in battle had to be not only sturdy and strong but steady enough to keep ranks under fire.

Above: domestic stallions used for selective breeding, in a painting by the 18th-century British painter George Stubbs.

tively bred for the sturdiness required to carry the warrior and the extremely heavy armor that protected both of them.

The Oriental breeds of which the Arab is a prototype are elegant horses which walk with heads and tails held high and ears pricked. Although small-bodied, these beautiful creatures are very fast and able to carry great weight. The first wild Arab horses roamed the rough deserts of the Yemen as early as 5000 B.C. Baz, the great-great-grandson of Noah, is alleged to have been the first to capture and tame one of these chestnut horses. A statuette discovered in Egypt and dating from 2000 B.C. shows a man on the back of a horse which closely resembles a modern Arab—the first evidence of people riding horses. Lightly built modern riding and racehorses owe their existence to three Arab stallions brought to England at the beginning of the 18th century and crossed with some of Britain's heavier breeds.

Selective breeding has produced many other domestic farm animals, including poultry or fowl—chickens, ducks, geese, and turkeys—and the pig. Modern domestic pigs are quite different from their wild ancestors, although early breeds bear a closer resemblance to the wild pigs from which they are descended. The Eurasian wild boar, for instance, an omnivorous, long-legged, long-snouted, heavy-shouldered, short-bodied creature covered with a coat of coarse bristles, is known to have existed in Neolithic times since its remains have been found with those of short-horned cattle and small sheep of that period. During Saxon times in Britain pigs were herded into oak forests for much of the year.

The modern domestic breeds of today are shorter-legged, fatter animals; they sport the typical curly tail and are usually only sparsely covered with hair. As with most domestic animals, their color is variable and often piebald. Today's pig breeds have been molded over the course of this century, and different types have been developed to answer varying meat requirements for bacon, pork joints, or the needs of the manufacturing trade. Meat is the most important product, and pigs have therefore been bred to mature faster, fatten more easily, and reproduce earlier and at a higher rate than those breeds domesticated for their muscular power or their hides.

Adapting to an Environment

Below: European wild boar (*Sus scrofa*) with young. The Eurasian wild boar (including the European, Indian, and Chinese species) usually grows to a length of 4 feet and sometimes as long as 6 feet. Its tail may be up to a foot long, and its height at the shoulder can be up to 3 feet. The boar may weigh up to 420 pounds and the sow up to 330 pounds. Its tusks may be a foot in total length, including the continually growing root. The body hairs are sparse bristles with some finer hairs interspersed; the tail is covered only with short hairs.

Below right: domestic pigs. The modern pig has been bred almost solely for its flesh and its fat. But its bristles are also used for making brushes, and the hide goes into sandals and other leather goods. Even the bones may be ground up for bone meal fertilizer. In the past the living animal was put to a variety of uses, too—for religious sacrifices, pulling carts, and detecting truffles. In ancient Egypt pigs were used for sowing corn, as their sharp hoofs made holes of the right depth for the seed to germinate. In medieval Britain pigs were trained as pointers and retrievers for illegal hunting or poaching, during the period when common people were forbidden to keep large hunting dogs.

Some domesticated animals—tamed camels, reindeer, and yak—lead lives very similar to those of their wild ancestors and still look very much like them. They have not altered greatly because their benefit to man lies in their suitability to their particular environment. The yak is supreme in the mountains of Asia, the reindeer is equipped for life on the Eurasian Arctic tundra, and the Arabian and Bactrian camels can withstand the harsh, arid environment of deserts where the climate includes extremes of both hot and cold. Domesticated reindeer retain their innate urge to migrate as the seasons change, and as a result their nomadic herders follow their movements and only round them up at intervals. Actually semidomesticated, they provide milk, meat, and hides, and some carry packs and draw sledges. They are herded in Lapland, northern Siberia, and in northern Canada (where they are called caribou).

The Arabian camel no longer exists in the wild, and indeed the two-humped Bactrian camel or dromedary faces the same plight. A few dozen individuals inhabit the remoter parts of Mongolia, though again it is doubtful whether they are truly wild or merely feral. In Australia herds of aggressive and bad-tempered Arabian camels, descendants of others which survived early explorations of the interior, have become feral.

Truly domesticated animals breed successfully in captivity and are entirely dependent on man for their survival. Most are completely different from their ancestors as a result of intense and careful selective breeding. However, the Asiatic or Indian elephant, *Elephas indicus*, is an example of a wild animal which can be tamed in early youth to perform tasks throughout its long adult life (they can live to be 70 years old). Although these elephants do breed in captivity it is common practice to capture and subdue wild ones for the logging trade. Once their spirit is broken they become extremely docile, doing astonishingly hard work for their drivers, the mahouts.

Above: domesticated llama (*Llama peruana*), member of the camel family, carrying a pack. It is still important as a beast of burden to the people living high up in the Andes of Peru. Llamas are particularly suited for life at these heights because their hemoglobin can take in more oxygen than that of other mammals, and their red corpuscles have a long lifespan—235 days as opposed to the human's 100. Aside from their use for transportation of goods weighing up to about 50 pounds, llamas provide meat, wool for clothing, hides for sandals, and fat for candles. The wool when braided is used for rope, and their dung is dried for fuel.

Left: a working elephant (*Elephas indicus*) moving logs in India. The elephant is the largest living land animal, and the smaller of its two surviving species, the Indian elephant, is also native to many parts of southeast Asia. The diet is entirely vegetarian and includes grass, foliage, branches of trees, and fruit.

The Farming of New Breeds

The cheetah is another species which has been domesticated successfully without imprinting the image of man on very young offspring. Until the last few decades cheetahs had never been bred in captivity, and it was adults that were caught and trained for coursing game in India. The process of taming these wild creatures was a type of imprinting, in that the adult was first weakened by being kept without food or water for a week, after which it was tied down under a thick sheet and beaten mercilessly. When the trainer then offered food, water, and caresses the cheetah learned to fix its future expectations on this person.

Until recently man has tended to avoid domesticating most other wild creatures. However, in the last few years animal breeders have begun to experiment with certain species, mainly because they are better adapted to their environment and the local vegetation than an introduced domestic breed. Rather than kill wild animals whose activities interfere with man's, as has happened with the red and gray kangaroos of Australia and the red deer of Europe, might it not be possible to farm them like sheep and cattle?

The red deer of Britain and Europe are better adapted than domestic breeds of sheep for a mountain habitat. Their diet is much the same as that of sheep—grass and young heather shoots—but in hard times they exist on what is available, including mosses, tree bark, and twigs. Sheep do not resort to these food sources. Red deer are also good potential meat animals, since

Below: Masai herdsmen with their cattle in Kenya. The pastoral Masai are fully nomadic and wander in bands throughout the year, subsisting almost entirely on the produce of their livestock, which also play an important part as the dowry or bride-price in the marriage contract.

they grow faster than young sheep in summer, although they do lose weight in winter even if sufficient food is available. Experiments in progress in Scotland seem to indicate that tameness can be passed from one generation to the next not only by hand-rearing young calves but by selecting particularly docile wild hinds for breeding; the characteristic for amiability may be inherited. However, there are many barriers both of principle and prejudice which so far block the raising of deer for food.

In comparison to the number of wild animal species the number of domesticated species is not very high. However, the total number of individuals is very high due to their value as a source of high-protein food for humans. Agricultural economists are now attempting to persuade people to turn to high-protein plant crops such as soya beans because they produce a very high protein yield for less investment of time and money than that required to raise domestic animals. Another extremely important factor is that in the near future there will not be enough land or grass to support the number of domestic animals required to supply protein to an expanding population.

At present domestic animals are the most important source of protein for omnivorous man. In 100 years they will no doubt be less important because man will have been obliged by then to change his diet; plant proteins will have supplanted animal proteins. Domesticated animals will then be kept for their other attributes as laborers and pets and not for their food value.

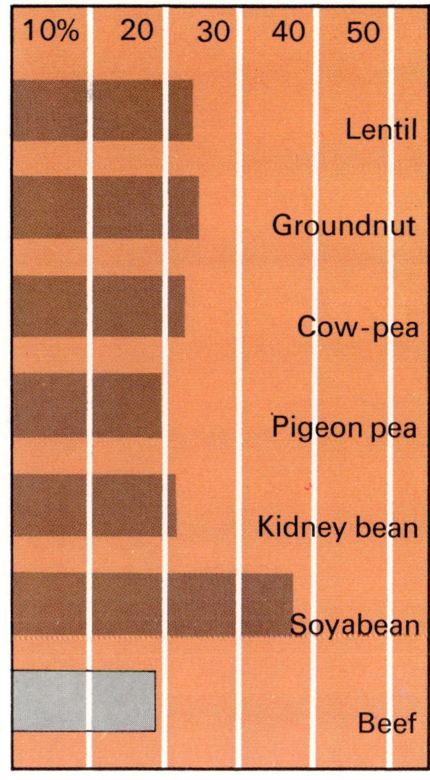

Above: comparison of the protein content in various legumes and in beef. The soyabean is even higher in its protein percentage than beef. Carbohydrate content is almost as high as in cereals (about 60 percent), and the fat content is generally low, except in soyabeans and peanuts.

Left: cows grouped in pens according to their grade of milk. Mechanization in the rearing of cows was just beginning in Britain in the 1890s. Since then many technical devices have been used to increase productivity, such as conveyor belt feeding systems and automatic milking parlors.

Chapter 12
The Social Insects

Most insects are solitary—they gather in groups only to mate or to devastate crops (in the case of locusts and caterpillars). Three types are, however, truly "social"—bees, wasps, and ants all live in vast, highly complex communities where specialists do particular jobs to insure the survival of the group as a whole. Dominating the insect world are the ants, which among themselves represent the three major phases of human development—hunting, herding, and farming. How do such tiny creatures manage their well-known feats of building and foraging? How does each nursemaid, cleaner, soldier, hunter or queen ant know what to do? How do social insects communicate among themselves? This chapter explains how patient observation has led to some answers, but other questions remain mysteries.

Insects are all around us in all parts of the world except the sea, from the insides of our houses to the tops of mountains and the ice of Arctic wastes. The smallest are tinier than some single-celled protozoans, while the largest ones are bigger than a mouse or shrew. Insects are the most successful of all classes of animals in the struggle for survival, due primarily to the ability of most of them to adapt to changing environmental conditions. In addition most insects are loners that meet other members of their kind briefly only to mate, although some gather to travel together (butterflies, moths, and locusts), others feed together in huge numbers (aphids or greenfly), and a few (such as ladybirds) congregate in hundreds to hibernate.

Yet there are three groups of insects that have evolved highly developed, complicated societies—termites, ants, and bees and wasps. Although at first glance termites look like ants, they are in fact given an order to themselves (Isoptera) and are more closely related to cockroaches. Ants, bees, and wasps are closely related members of another order, the Hymenoptera.

These insect groups live within a wide variety of social systems. A single colony can contain over a million individuals working in harmony, and ethologists and entomologists have always been fascinated by their smooth coordination. Nowhere else in the

Opposite: nest of the bumblebee (*Bombus agrorum*), a genus of social insects found all over the world. Most species live in the tropical or subtropical zones, but they can also be found in places as far apart as Arctic Canada and Tierra del Fuego. They are not native to Australasia but were introduced there when the early settlers found that none of the local bees could pollinate the imported red clover plant. For a nest a new queen bumblebee may choose to take over the abandoned nest of a fieldmouse, a vole, or a hedgehog. Or she might select a disused bird's nest, a thatched roof, a bale of hay, or even a discarded mattress. Bumblebees, which are in a sense vital to the pollination of crops, are actually becoming scarcer due to modern farming methods, which have cut down on the hedgerows and neglected corners of fields in which they prefer to nest.

The Amazing Leaf-Cutting Bee

Below: piece of bamboo split open to show the nest of a leafcutter bee (*Megachile centuncularis*) inside. This nest consists of three cells which hold the larvae and their food, and it is sealed by a thick plug of leaf at the open end.

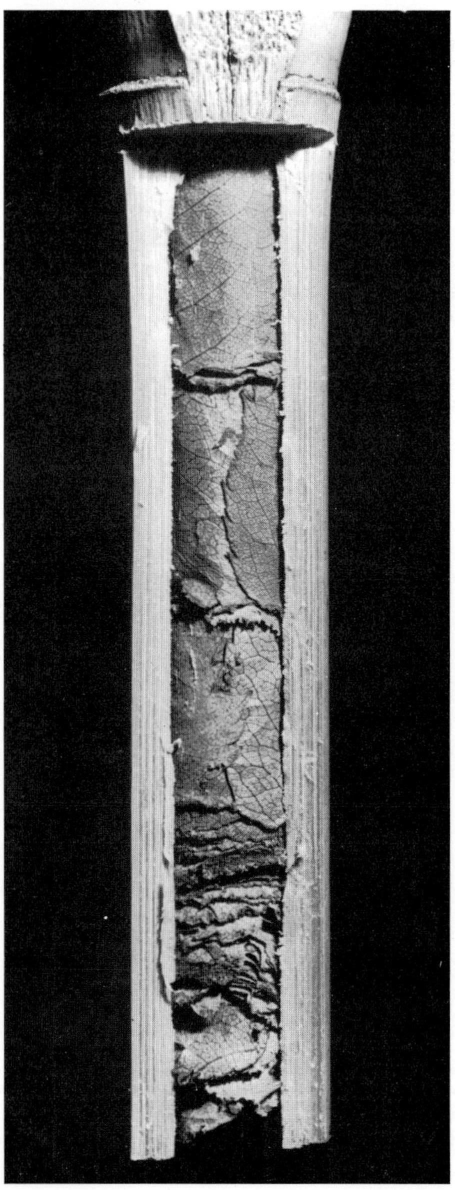

animal kingdom has a similar level of organization evolved, and this uniqueness only adds to the mystery.

In most large social groups of mammals or birds the majority of individuals are unrelated. By contrast an insect colony, whether of bees, ants, or termites, consists primarily of the daughters of a single parent, the queen. She is an offspring-producing powerhouse, and in many cases it is her whole life. Most of her eggs hatch into daughters which, though genetically female, are unable to reproduce; they are the workers of the colony. Males exist solely to mate with virgin queens.

Because of their close family relationship the individuals of an insect society inherit the same genes and instinctive behavior patterns. This lack of genetic variation is the source of the amazingly close cooperation between members of the colony. Antagonism does occur among members of the sexual broods which will eventually establish competing colonies. Research has shown that males in colonies of army ants of the genus *Eciton* compete with and cannibalize each other while the workers do not.

Most zoologists agree that the term *social insect* describes those in which the generations overlap, live together, and cooperate in their work. A division of labor and jobs is usually involved, delineated by a caste system in which individuals of the same species (and sex) are different in structure and function.

Of the three social groups, the social organization of bees and wasps is the simplest. In fact many bees and wasps lead solitary lives apart from mating time. But they perform tasks of parental care such as nest-building, and in fact their behavior patterns are similar to those of social honeybees except that each female acts as both queen and worker in order to raise her young. Solitary bees do not make wax as do their social relatives. Instead the nest is made of plant material and situated in a hole in the ground or cavity of a tree or plant; each species has its own preferred nesting site. The same plant material is used to make the cells of the nest, and food—usually pollen soaked in honey, or *beebread*—is put into each cell. The female seals off her nest once she has laid an egg in each cell and made sure that there is enough beebread to feed each larva after it hatches. In some species the female dies before her offspring mature while in others she provides a certain amount of parental care.

One of the best known solitary bees is the leaf-cutting bee of the family Megachilidae. Gardeners are familiar with this bee for the neat oval holes it cuts from the leaves or flower petals of rose bushes and other plants. The bee rolls its pieces of leaf around masses of pollen, grouping several together to form cells which look like a group of sewing thimbles. She lays an egg in each thimble and then seals it off with another piece of leaf shaped in an almost perfect circle to fit the thimble.

There are three main types of truly social bees, all belonging to the subfamily Apinae: bumblebees, honeybees, and the Euglossini, a small group of tropical stingless bees. Each species contains three castes: workers (undeveloped females); queens; and drones (males which exist only to fertilize the queen's eggs).

Most bumblebees live in the temperate areas of both the northern and southern hemispheres, but some are found surprisingly close to the poles—from Greenland in the north to

Above: the leafcutter bee at work on a section of leaf. A solitary bee as opposed to the social honeybees, this creature uses natural materials to erect nesting structures suitable for its larvae's development.

Tierra del Fuego in South America. None of these societies survive the winter but must begin a new colony each spring. Only the young queen bees survive through the winter by hibernating underground. They emerge in the spring with the first blossoms, when they hunt for a suitable site to found a new nest. This may be an old mole nest or perhaps a dry tussock of grass. The queen constructs a honey pot in a small cavity with wax exuded from her abdomen. She fashions another wax cell for the eggs she then lays, and after three to four weeks her first small batch of workers emerge.

These young workers soon begin their task of gathering nectar and pollen, building more cells to store it, and caring for the new batches of eggs the queen lays. She is now able to concentrate on her main job of egg-laying. The colony grows quickly until it reaches its maximum size of just under a thousand individuals. In the summer male drones and young queens are produced, the first from unfertilized eggs and the second from larvae that are fed on a special substance called "royal jelly" which is secreted from the salivary glands of 6- to 14-day-old worker bees. Whether or not they produce the royal jelly seems to depend on how much contact each worker has with the queen. Workers in close proximity to her are prevented from making queen brood cells or producing the jelly by a secretion from her body called

Household Duties of the Honeybee

Below: bumblebee (*Bombus agrorum*), larger than a honeybee and covered with stiff orange, yellow, or red hairs. The bumblebee has a sting it can use to inject venom into the body of an enemy, but it is not barbed like the honeybee's and so can be withdrawn. The sting itself is a modified *ovipositor*, a tubelike organ used by other insects to deposit their eggs. Though bumblebees are social insects, their communal life is not as well developed as that of other bees. There are fewer workers in a colony, and they all die by wintertime.

the "queen substance." This is distributed to the workers closest to the queen, but with advancing age the quantity she can produce lessens. Those workers who do not receive the substance then begin to produce new young queens by transfering some eggs to queen cells and feeding the hatched larvae on royal jelly. With the approach of winter the old queen and the workers die, but the young queens survive to found new colonies.

An interesting point about bumblebees is that their particularly long tongues enable them to reach nectar produced in the base of flowers like the red clover, which have very long tubes. When first introduced into New Zealand for its nitrogen-producing qualities clover did not thrive because its flowers could not be fertilized by the native bees or previously introduced honeybees. After a number of bumblebees were introduced from Europe the problem was solved. Honeybees originated in the Old World but the domesticated species, *Apis mellifera*, has been taken around the world to be used for controlled honey production.

Honeybees have the same basic society as bumblebees except that they are able to keep the colony alive through winter by storing enough food to last until spring. Extra food is stored in some of the wax cells, and it is this honey which beekeepers harvest from domesticated honeybee colonies. A flourishing hive can contain between 50,000 and 80,000 members, mostly workers. There is only one queen to each colony and a mere few hundred male drones.

The three castes of honeybees are more easily recognized than in bumblebees. The queen is the longest and largest of the three. The drone is a solid, chunky-looking insect with very large eyes, while the more compact workers have smaller eyes. A queen honeybee is ready to lay eggs within three or four days of mating. The workers busily prepare brood cells in a specific part of the nest where the temperature stays at around 90°F. Wherever the queen finds an empty brood cell she deposits an egg. She never leaves the hive once she has begun her tedious life of egg-laying—she is fed on a *broodfood* secreted from special glands by the workers. It is different in composition both from ordinary worker food and beebread, but how it is produced is still a mystery.

Both the queen and the workers lay eggs, but only the queen's are fertilized by the drones, and these eggs develop into females. The eggs of the workers hatch as males. For three to five years the queen lays egg after egg in order to keep up the necessary numbers of worker bees, because although a honeybee community cannot survive without an active queen, the backbone of the colony is the workers.

Worker honeybees perform many duties, depending on the age of the individual bee and the specific needs of the community. Workers' lives are relatively short, which is why the queen must continually replenish the supply. Those which hatch in autumn survive through the winter and early spring until the new summer bees take over.

As soon as a young worker bee emerges from a wax cell it gets on with the basic "household" duties of the hive. It solicits food from the jaws of older workers and helps itself to the honey and pollen of the storage cells. It is continually looking for jobs to

Above: worker honeybees receiving the "queen substance" from the queen bee (center). This and other substances are passed on to all members of the colony to make sure they are aware that their queen is well and present in the hive, and so is a form of communication between social insects. As the queen ages she produces less of this substance, and as a result new young queens are produced by some worker bees and fed on royal jelly. This insures the founding of new colonies and the survival of the species.

Left: worker honeybee with its pollen baskets, special sacs situated on its third (rear) pair of legs. Each pair of legs does a different job. The front legs, the shortest pair, have branched feathery hairs for collecting pollen. A special joint enables the bee to clean both its eyes and its antennae. Stiff hairs on the middle legs brush pollen from the thorax and front legs. The rear legs scrape pollen into their pollen baskets.

perform such as cleaning out recently emptied brood cells. If a crisis arises, as when a predator breaks into the hive to take the honey, these young workers help to repel them. Only workers and queens have stings—drones are defenseless. Young workers also help to feed older larvae with pollen and honey, but they cannot care for young larvae until the broodfood glands have matured at the age of around 12 days. A few days later the wax-producing glands are operating so that they can build and repair wax combs and cells. Another week is spent storing honey and pollen brought in by older foraging bees. At this time the female makes her first exploratory trips away from the hive—she has not yet used her wings for flying and so practices while perfecting her orientation techniques.

At the age of three weeks a female worker enters the food-collecting period of her life. She leaves the hive during the day to search for pollen and nectar which she brings back to the hive in *pollen baskets*, special sacs situated on a third pair of legs. After three weeks of work (for summer bees) the worker dies.

The behavioral patterns of honeybees are sufficiently flexible that if too few older workers are available for food-gathering, young ones join the foraging gang, and if too few young bees are available to take care of all the household duties, older ones help out.

Male drones, although they are completely cared for by industrious females and their sole function is to mate with emerging queens, do not have an easy life. Theirs is a precarious existence because it is completely controlled by workers. If there are too many males in the hive, the females destroy drone larvae by dragging them out of their cells. They tear down the drone cells and rebuild them as worker cells.

What triggers off a swarming to establish a new colony is not known, but the necessary stimulus may come from overcrowded conditions. New queen larvae are usually developing just before a swarm takes place, but there are exceptions. Usually about half the population of the hive takes off—an impressive sight as over 30,000 bees can be involved. Then they settle while scouts seek out a suitable nesting site. When they report back the swarm moves to the chosen site and prepares to establish a new colony. They can survive for a few days without foraging in order to get on with the important task of building cells because many of the workers bring with them supplies of nectar, which are shared among all the bees.

Bee societies are simple compared with those of ants and termites. These colonies are often graded into numerous worker castes ranging from the smallest creatures, whose job it is to provide food for the colony, to the largest members that defend it. The larger ones are usually called soldiers and have formidable jaws, or *mandibles*, that can bite as well as sting.

There are over 8000 ant species of the family Formicidae which occur worldwide but are most common in the tropics. The early stages in the life of a queen ant are similar to those of bees. She takes to the air and mates with a winged male, after which she works at a nest, usually underground. She rubs off her now useless wings and uses the wing muscles as a food reserve as she begins to lay eggs. After hatching the larvae are fed on the

What Triggers a Swarming?

Opposite: a swarm of bees on an apple tree. When searching for a new home honeybees settle in a large cluster while scouts fly farther afield. The swarm is made up of large numbers of workers, a few drones, and an extra queen produced in the formerly overcrowded hive. While settled they can be persuaded to move into artificial quarters merely by shaking the swarm, with its queen, into a suitable container. The next stage is an egg-laying marathon.

Below: a royal birth—a queen honeybee struggles out of her cell. From her body the queen secretes a substance whose presence or absence controls the behavior of her great horde of daughter-workers, who shoulder the burden of supplying the colony with food, shelter, and protection.

Above: winged male black ants (*Lasius niger*), which mate with a queen ant on a nuptial mating flight, after which she rubs off her wings, works at starting a nest, and begins to lay eggs.

queen's saliva for a few days until they reach the pupa stage. The ants that develop from these are relatively small and are often called *minims*; they break out of the nest to forage for food. The queen settles down to laying eggs, which she will continue to do for the next 10 years or so, drawing on reserves of sperm obtained from the single mating.

One major difference between worker ants and bees is their lifespan. Worker ants can live for five or six years in most species, and some live much longer. Their colony is also generally much larger, although some of the simpler ones may contain as few as 50 individuals. A large colony can reach over a million inhabitants as with certain army ants.

Inside an ant nest are several chambers. The most active is the one where the queen is fed and groomed by the workers as she lays her eggs. Other workers take the eggs to brood chambers in the warmest parts of the nest. On hatching the young ants are carried to a nursery, and there are other chambers for food storage and even for refuse.

The most notorious ants are the army ants (*Eciton*), found in tropical America. Various species are called soldier ants, driver ants, legionary ants, foraging ants, and visiting ants. These nomads prefer to make a base camp while the queen lays her eggs and then move on. They never build a permanent nest. Carnivores, they march along in their hundreds of thousands devouring the insects and other invertebrates in their path. They play an important part in keeping the balance of nature steady within their environment. They can clear a house of insects, mites, lizards, and so on in a very short time. Stories of people being

Flesh-Eating Army Ants

stripped to the bone in minutes are greatly exaggerated, although the ants certainly do bite.

Army ants have evolved several adaptations in their social structure and reproductive habits to fit in with their traveling. Research has shown that most species go through two major phases of activity: a sedentary or "statary" phase, during which the ants forage on a small scale from a central bivouac inside a hollow log, and a nomadic phase, during which the entire colony moves at night while camping for a day or two at a time in a more temporary bivouac under a log. Each statary phase lasts between 17 and 22 days and during this time the queen lays about 180 eggs an hour. The larvae from the previous egg-laying cycle are now pupae which hatch, and when the workers emerge from their cocoons the statary phase ends. The ants then begin their nomadic phase, during which the young larvae grow and require abundant food. They are carried each day to a new bivouac. When the ants produce a brood of males the nomadic phase lasts only 8 to 13 days (instead of the usual 16 to 18 days) because male larvae develop much faster than worker larvae. Winged sexual males are produced only once a year, and when mature they swarm through the tropical forests in search of virgin females.

There is only one queen in a society of army ants, but the workers are divided into several castes including major workers or soldiers and several smaller minors. A typical army ant colony consists of 100,000 to 500,000 workers; 50,000 to 200,000 eggs; and as many as 60,000 larvae. Larger colonies may contain 1.5 million adult workers.

Left: army ants (genus *Eciton*) on the march. Each ant is often over an inch long. The queen is wingless, but the males are winged and distinctly wasplike in appearance. There may be more than one kind of worker, each physically adapted to do a particular job. Their moving columns may stretch for yards, and anything in their path that is too slow to escape may be attacked. A tethered horse will simply be eaten alive and left a skeleton. Army ants feed by cutting up and rending the victim on the spot and carrying the pieces back to their temporary nest. They also fill their crops with juice and pulp. An average-sized colony may require half a gallon of animal food daily, purely for subsistence.

Above: matabele (safari) ants (*Megaponera foetens*) of South Africa returning from a raid on a termite nest carrying numerous termites in their jaws. Each ant is about half an inch long.

In an army ant colony only about six new queens are produced during the annual sexual phase. They are ready to mate at once but the 1000 to 2000 males that emerge a day or two later cannot mate for several days. The males of certain species must make a dispersal flight first, but in any case due to this timing the males of one colony are unable to mate with virgin queens of the same colony. A queen mates with a male of another colony, thus allowing cross-fertilization, and once this has taken place the queen takes part of her parent colony with her to establish her own group.

Army ants have not acquired their name simply for their marching but also because of their raids. Some species form narrow columns to capture the larvae of other ants and wasps from their nests. Others form a swarm which can be as wide as 65 feet across at the front, flushing insects from the forest floor as they advance. Raids are usually made shortly after dawn, and they become larger during the final days of the nomadic phase because there are so many developing larvae that need food.

During their nomadic phase the colony starts moving to a new bivouac in early afternoon, not stopping until late in the evening. Along the 500 to 650 feet covered in a day the ants carry larvae and booty from the previous nesting place to the new one. Guard workers are scattered along the column to attack anything trying to cross the trail of the emigrating ants. By the time darkness falls three-quarters of the ants have left the old bivouac, and the queen sets off for the new nesting site. The main advantage of their nomadic habits is that they can greatly extend their foraging area and thus obtain more food so that very large colonies can be supported without the risk of starvation.

Other ant species have evolved the skill of farming. Some grow crops for food while others raise insects for the secretions they emit. Some species even capture other ants and keep them as slaves to work for the colony.

Fungus Grown by Farmer Ants

The 200 or so species of crop-raising ants (genus *Atta*) of the New World are usually called leaf-cutters. They cut out pieces of leaves and, holding them high in their jaws, carry them back to their nests; the sight of them marching along has given rise to their other name of parasol ants. Leaf-cutters nest in gigantic underground colonies which may contain over 500,000 ants and more than one queen. Worker ants deposit the leaves in one of the special chambers within the nest where they are chewed to form a compost in which the ants grow a special kind of fungus. These fungi, different from the familiar mushrooms and toadstools, have evolved along with the ants, which meticulously weed out foreign fungal growth from the "beds." With pruning, rounded knobs called *bromatia* are produced on the fungi, and it is on these that the ants feed.

When a young queen sets out to establish a new colony she takes along a little of the fungus of her particular species stored in a small cavity near her gullet. After she has mated and dug the beginnings of her new nest she regurgitates the fungus. The queen starts a new garden and tends it until the first workers have developed enough to take over.

Below: leaf cutter or parasol ants (genus *Atta*) of the tropics and subtropics of the New World, carrying pieces of leaves in their jaws. They slice the leaves up into sections larger than their own bodies and can then be seen in large numbers transporting them along well-defined trails back to their nest. The ants themselves are large and long-legged with spines on their bodies. Inside special chambers within the nest worker ants chew the leaves to make a compost on which a special kind of fungus is grown. Each species of leaf cutter ant grows its own particular kind of fungus and weeds out the spores of any other type.

Ants That Milk Greenfly "Cows"

Harvester ants (genus *Pogonomyrmex*) are seed-gatherers, whose activities have been compared to pre-agricultural man. Collected seeds are stored in chambers called granaries for those periods when little food is available in the ants' arid, semidesert or desert habitat.

Greenfly, scale insects, plant lice, and even the caterpillars of some butterflies all produce a sweet secretion from plant sap called *honeydew*. Many ant species collect this substance for food, and some actually "persuade" the insects to provide it for them. Gently stroking with its antennae a feeding greenfly whose body is gorged with honeydew, the ant milks the greenfly which exudes droplets of the sap. When the ant has lapped up as much honeydew as it can hold it returns to the nest. Continuous milking results in the build-up of the greenfly's output the way a cow's milk production is increased with domestication. The ants spray formic acid on potential enemies of their "cows," and if this does not deter them the ants pick up the greenfly and transport them to a safer place. When protected the greenfly are able to greatly increase their numbers and therefore the ants' food supply.

In some species the "cows" are moved from one plant to another so that fresh sap is always available to the feeding greenfly. Some even tend greenfly eggs in their underground nests over the winter. The corn root aphid, *Anuraphis maidiradicis*, of North America is cared for by the ant species *Lasius americana*. In spring the newly hatched corn root aphids are carried out to feed on young grass near the nest and are moved to corn plants as they grow, causing a great deal of damage to the corn crop.

In very dry desert regions the bodies of some ants have become adapted to storing honey. These are the honey or honeypot ants of the genus *Myrmecocystus* which are found in North America, southern Africa, and Australia. A special subcaste, called repletes, swallow the honeydew brought in by returning workers. As the abdomen gradually swells the narrow black bars of the body segments are forced apart and between them, attached to the thorax and head of the ant, round globes of honey are formed. When fully swollen the ants hang from the roofs of underground chambers. When no food can be foraged from the desert the workers solicit the stored honey; gradually the abdomens of the honeypot ants gets smaller, but because their swollen abdomens

Above: four ants tending their greenfly "cows" as they feed on plant sap. Many ant species collect the honeydew produced by the greenfly and use it for their own food.

Right: harvester ant (*Messor barbara*) carrying seed back to its nest chamber. The granaries are used to supply the ants with food during periods when they might otherwise starve.

do not contract to their former shape, these ants are doomed to die of exhaustion once their task is fulfilled. Inhabitants of the Mexican countryside and the Aborigines of Australia consider honeypot ants a delicacy.

Robber ants, *Formica sanguinea*, are notorious for enslaving other species. They send raiding parties into the nests of other, often closely related species such as the black ant (*F. fusca*). After capturing as many pupae as possible they return to their own nests. If the black ant workers fight back they usually suffer worse casualties. A few of the captured larvae are eaten but most are allowed to hatch, after which they take on tasks within the nest and are apparently accorded equal status with worker robber ants. The warrior ants of the Amazon, another slave-making species, depend on their slaves for their own survival. The jaws

Below: interior view of the home nest of a colony of Amazon warrior ants (*Polyergus rufescens*), here depicted in light color. They have made slaves of the dark-colored black ants (*Formica fusca*) by raiding their colony and taking away cocoons containing the pupae of *F. fusca* workers. After emerging from their cocoons these perform all the household labor of the Amazon colony. At top center one of the slaves brings a fly wing into the nest for food. Other slave workers care for the small eggs, grublike larvae, and cocoon-enclosed pupae of their captors. In order to eat, the Amazon ants must beg slave workers to regurgitate liquid droplets for them (lower left).

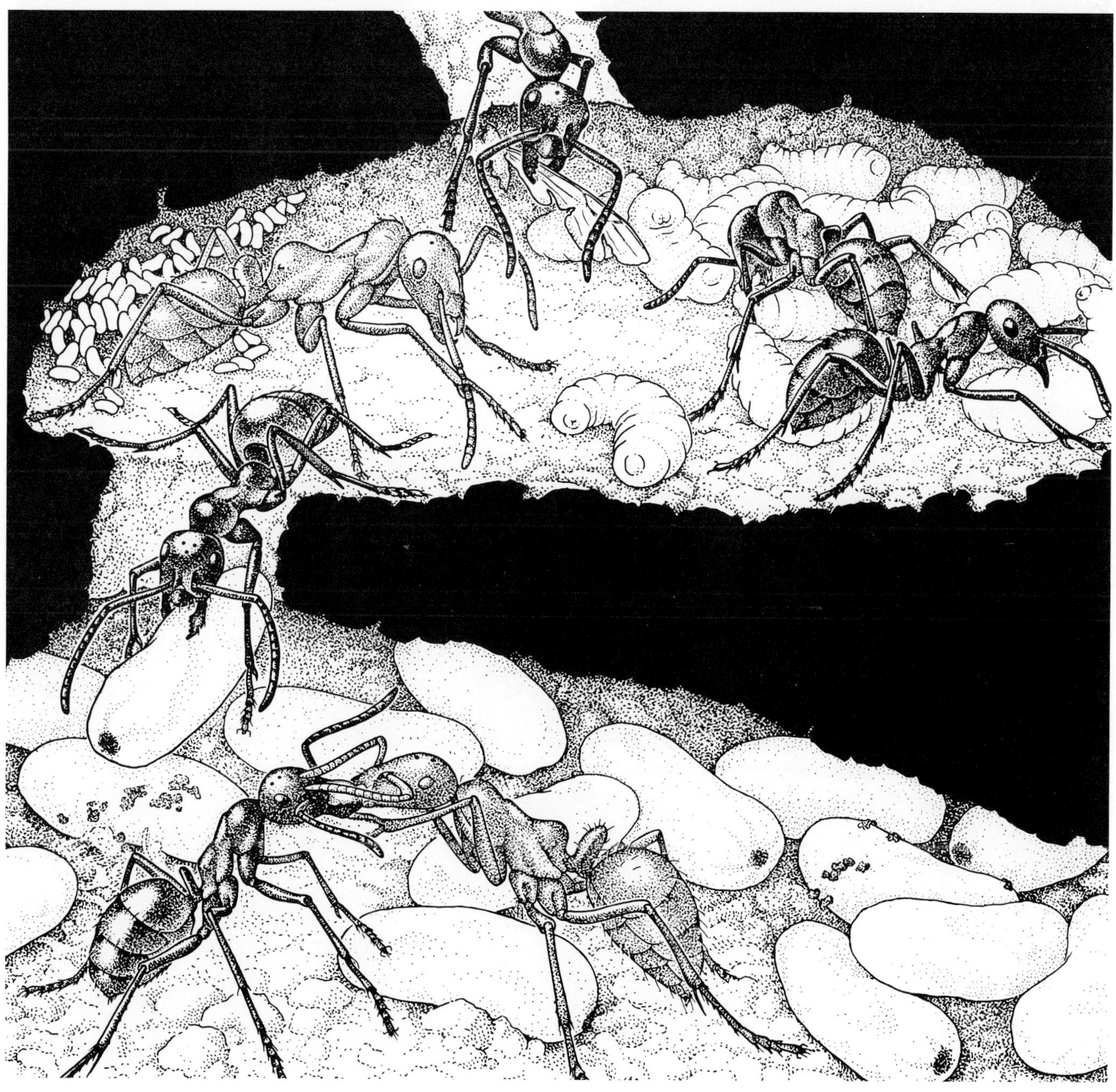

of warrior workers are designed for defense or attack and are useless for tending eggs and larvae. Thus the slaves are essential for the care of warrior young.

The third group of social insects are the termites, often incorrectly called "white ants." Superficially they resemble ants, but termites have evolved independently and their organization is far more complex than that of an ant colony. The 1700 species range in size from the giant African termites whose queen can reach 5 inches in length to some tiny species just over an eighth of an inch long. Most species live in the tropics but a few are found in southern Europe and the United States.

Although termites, like ants, have a distinctive caste system, their life histories are very different. The distinct larval and pupal stages of ants do not occur in termites. Instead, from the eggs hatch tiny termites, miniature replicas of the adults. As they grow they do not change their structure but simply become larger until they reach adult size. At first fed by adult workers, the young soon learn to feed themselves.

At the head of the termite colony is the queen, much bigger than any other member. Unlike ant queens who store enough sperm from their one and only mating to last their lifetime, the queen termite founds the nest with her mate and together they dominate the colony. During their lives they mate at intervals and are always attended and fed by the workers.

Another difference between ant and termite castes is that termite workers can be of either sex, although neither is capable of mating or reproducing. There are many variations depending on the species of termite, but generally the soldiers are found in three forms—major, intermediate, and minor. All soldiers perform the same functions but how they do them depends on the physical equipment of the individual soldier. Some have particularly large heads which they use to block the openings of the nest to intruders. Some have knifelike mandibles that can snap and bite, others can inflict blows, and still others spray corrosive chemicals.

The more primitive build their nests or colonies in wood in which they excavate chambers; in most cases the wood provides

Termites-Mites or Monsters?

Opposite: a honeypot ant (*Myrmecocystus mexicanus*) with a swollen abdomen full of stored honeydew. The workers which remain in the nest and act as living storage vessels are called repletes. The so-called "honey" is a sugary solution obtained by the ants from aphids and from the secretion of a *gall*, an abnormal growth, which forms on small oak trees.

Left: termite mounds. More primitive species build their nests in excavated chambers in wood, but African and Australian species construct huge termitaria covered with an insulating layer that functions like an outer shell of reinforced concrete. In some regions they are so numerous that these gigantic structures determine the character of the landscape. Different-sized cells house the different castes of termite, from the royal cell of the queen and her mate to interior chambers occupied exclusively by eggs and larvae.

Below: 2.5-inch-long queen of the termite species *Macrotermes bellicosus* of southern Africa being tended by workers. The queen's abdomen is monstrously enlarged because of the development of her ovaries and can become as large as 6 inches long by 1.5 inches wide. She may produce as many as 30,000 eggs in a day. Eggs leaving the tip of the abdomen are taken over by the workers. The king usually sits next to his queen, who is also guarded by soldiers (with larger heads).

them with food as well. A ventilation system regulates the oxygen supply to the nest, gets rid of stale air, and keeps the temperature of the nest constant. A species on the Ivory Coast of Africa constructs a huge nest or termitaria with an insulating cover layer. Like harvester ants, these termites grow fungal gardens in the center which receive fresh air through channels leading up from a cellar cavity. Stale air is expelled at the surface entrance of the nest.

At certain times of the year sexual winged males and virgin females develop and fly away from the termitaria to mate. Each pair then lands at a suitable spot, their wings fall off, and the new king and queen begin to dig a new nest. Eggs laid by the queen hatch into nymphs, some of which grow into workers that maintain the colony while others develop into soldiers which defend the nest. It is thought that growth-inhibiting hormones determine the development of the various castes. Until the nymph workers are old enough to forage for food the royal pair survive on stored fat and their now useless flight muscles. Sometimes they also cannibalize some of the eggs or nymphs.

Of all the social insects termites are usually the longest lived. The royal pair live from 15 to 50 years depending on the species, and there is evidence of one queen living for over a century. Workers and soldiers live for about four years.

Termites are not often found foraging above ground like ants. They prefer to remain under cover and tunnel their way to food through rotting fallen wood, living trees, or man-made timber structures. The surface of the wood is left intact while the inner wood is carefully hollowed out over a period of time. In the end the structure is liable to crumble at a touch.

The Riddle of How Termites Communicate

Left: termite nymphs (*Macrotermes bellicosus*) a fifth of an inch long tending tiny white fungi in their nest. These fungal gardens are in the center of the colony and receive fresh air through channels leading out of the termitarium. The fungi are used as food for the entire colony.

How the activities of hundreds, even hundreds of thousands, of individuals can be so well coordinated that their colonies work efficiently has been the subject of intensive research. The creatures do not continually touch one another, and they make very little noise—so how do they communicate? It appears that odors, visual displays, certain sounds, and intermittent touching form an elaborate system of behavioral stimuli.

Smells are particularly important. Hormonal substances called *pheromones* are passed between individuals of a colony and stimulate particular behavioral or physiological responses from the recipient. For example, ants move along a particular course because they follow a trail of odorous substances which have been left by each ant preceding them. Certain ants can release warning pheromones that attract hundreds of workers and soldiers when the colony is in danger.

In addition to reacting to stimuli which trigger off innate behavioral patterns, social insects also learn by trial and error. The caressing movements of their antennae by which certain ant species "milk" their aphids may be learned. Ants learn the quickest, straightest route to a food source and waste less time on detours or backtracking with each trip. This ability has been demonstrated in experiments in which ants adeptly master mazes. Ants also orient themselves by the sun. However, we have as yet only scratched the surface, and the mystery of the lives of social insects remains to a great extent unsolved.

Chapter 13
Fragile Networks

All living things exist within a complex interplay of organism and environment, including both living and nonliving elements. What must man, as the most powerful species within that interplay, understand about how organisms affect each other? What must we know about the balance of nature in order to deal intelligently with the effects of our own activities? What lessons have we learned from such tragedies as Minamata? Is it possible to manipulate the environment successfully? What is the environmental cost of modern technology, and can that cost be accommodated within the structure of an industrialized society? Can man ever learn to live contentedly within the fragile networks of the planet earth? What will happen if he doesn't?

The whole of nature can be compared to a complex household in which everyone has a specific job, including the nonliving structure of the building itself. This concept has given rise to the word "ecology" from the Greek word *oikos* (house), and since the late 1960s it has become a subject of international interest. We have begun to realize that man, as the dominant species on this planet, has more influence over nature than any other living creature, and that his influence is increasing all the time. Recently it has become obvious that our power of interfering with nature, of overriding its dictums and altering its routines, has overtaken our understanding of how nature works. The result is the despoliation of our own home, the planet earth.

Modern science tends to separate knowledge into classifiable branches, and ecology overlaps several of these—chemistry, physics, geography, meteorology, and biology, to name a few. The subtle interplay between living and nonliving aspects of an environment is thus more difficult to understand, particularly for the Western mind, than would a straightforward cause-and-effect problem treated in isolation.

The plant and animal populations of a particular area, including each individual of the various communities that exist there, added to the nonliving environment which supports them, make up an *ecosystem*. This is a useful working unit for study

Opposite: vast areas of the world were once covered with forest, their floors carpeted with rich, fertile soil. Man has steadily drawn upon this valuable natural resource by felling trees for building materials and fuel and using the soil for crops or covering it over with roads or housing. This has meant not only the loss of habitat for many animal species, some of which may face extinction. It also means loss of the atmospheric oxygen that was once produced by the large concentrations of green plants.

Photosynthesis and Food Chains

because it shows its own internal patterns of relationships, but the boundaries are not always easy to define. One ecosystem merges into another just as the activities of one species merge into those of another. The important point is that an ecosystem includes *all* the organisms in it, from the bacteria in the soil to the birds and insects in the air, as well as *all* the factors of the non-living environment—nutrients, temperature, wind, relative humidity, light intensity, soil type, geographical position, and so on. Each of these factors has some effect on the living organisms and some also have an effect on each other.

Every household needs an income, and in nature this takes the form of energy, nearly all of which comes originally from the sun as light radiation. The flow of energy through an ecosystem requires a complex system of converting the light—since neither man nor any other animal can simply swallow sunlight—into a more usable form. Green plants are the converters, or *primary producers*—they make their own food by a process called *photosynthesis* in which they utilize the energy of sunlight through a material called *chlorophyll* to combine carbon dioxide and water into *glucose*. This sugar is a source of abundant chemical energy, and unlike sunlight it can be stored for use when needed. It is the main raw material from which all other matter in living organisms is made.

All life forms other than green plants—which include algae, mosses, ferns, flowering plants such as grasses, shrubs, and trees, and a few species of bacteria—are, in one way or another, consumers. Primary consumers are those that feed on green plants. On land these include not only *herbivores* such as grazing animals but also a multitude of microorganisms such as most bacteria and fungi that live on the remains of dead plants. In the sea the primary consumers include the host of small crustaceans that live on small, free-growing plants near the surface. The plants are called *phytoplankton* and the animals are *zooplankton*.

Right: horse chestnut leaves, which present the maximum area of photosynthetic tissue to the light. Their green color comes from the chlorophyll they contain—as a green pigment it traps red and blue light from the sun's rays and reflects the green and yellow light. Not only do the broad flat leaves trap sunlight, but they also gather carbon dioxide. Hundreds of tiny pores called *stomata* pierce the surface, and through them gases move between the cells and the atmosphere. Leaflike structures are essential for photosynthesizing land plants which must expose their chlorophyll to the light.

Primary consumers are eaten by carnivorous *secondary consumers*, which in turn may be eaten by *tertiary consumers* in the form of larger or fiercer carnivores. Preying alike on all these types are the *omnivores*, which eat both plants and animals. Omnivores are common in nature, for even the fiercest carnivorous animal occasionally supplements its diet with plant foods. The outstanding omnivore, of course, is man, who has a wider-ranging diet than any other animal. Finally, at the end of the line, microorganisms in the soil and in the sea break down the remains of dead animals, so making further nutrients available for other life forms.

The natural arrangement of who-eats-what is commonly called a *food chain*; actually, food web would be a better term because some species of plants and animals form a link between two or even more food chains. In any ecosystem there may be hundreds or thousands of separate food chains. The very fact that it *is* a community, and not a completely random assortment of organisms, means that each stage in a food chain is of critical importance to the equilibrium of the ecosystem as a whole.

Food chains are concerned with the transfer of energy from one organism (the consumed) to another (the consumer). But each stage of every food chain involves the conversion of energy into a form that can be used by the particular consumer for growth, for repair and maintenance, for physical activity, and for simple existence. The conversion process always means wastage of a certain proportion of the energy in the form of unusable heat. The energy-converting efficiency of organisms varies greatly—only between 1 and 20 percent of the light energy absorbed by plants is translated into glucose, and roughly 10 percent of the energy locked up in plant foods is available to the herbivore that eats it. Through the various conversion stages of each food chain all the energy originally derived from sunlight is gradually used up. But it is not destroyed—it is turned into unex-

Below: herd of grazing cows, primary consumers of green plants such as grasses. Herbivores provide a link between the energy producers and the carnivorous secondary consumers.

Indigestible Plant Food

Right: grazing cows eating away cliffs to obtain necessary supplies of salt from the soil. Herbivores in hot climates need more salt than can be provided by their vegetable diet. Carnivores may get enough salt from the flesh and blood of the herbivores they eat.

Below: lioness with her cubs after the kill. The lion is one of the largest living land-based predators, and it can survive only where there are many large plant-eating animals (primary consumers). Lionesses do more hunting than males, and they must range over a huge area if they are to find sufficient food.

ploitable heat, which is why life on earth requires the continuous transformation of light energy into glucose on a massive scale.

Plants need more than the raw materials of photosynthesis in order to build and maintain their own structures. Many of the molecules that form the bodies of plants, and that are vital to their functioning, contain other elements such as magnesium, nitrogen, and sulfur. Most photosynthetic land plants absorb these extra materials from the soil through their extensive root systems. Although nitrogen gas is a major component of the earth's atmosphere, green plants cannot draw nitrogen directly from the air but must take it either from nitrate salts in the soil or from bacteria that grow in their roots. These bacteria are able to capture inert nitrogen from the air and convert (or *fix*) it into ammonium salts and nitrates, compounds which green plants can utilize. Floating water plants absorb minerals directly from their fluid surroundings. For this reason the growth of phytoplankton is often inhibited by a low concentration of necessary minerals in the upper layers of the sea and of lakes.

Once plant food has been taken into the body of the primary consumer it must be digested before it can be used. Leaf cells, like all plant cells, are surrounded by a tough, supporting material called *cellulose*, which cannot be easily digested by most animals because they do not produce the right enzymes to break it up. To get around this problem, leaf-eating animals usually grind their food into tiny pieces, thereby fragmenting many of the leaf cells and making their contents available for digestion. Mammals have teeth with complex grinding surfaces, but insects and a few crustaceans have structures within their guts that perform the same function. The guts of some herbivores harbor minute bacteria and *protozoa* (single-celled animals) which manufacture the enzymes that digest cellulose.

Land plants do not always meet all the requirements of primary consumers, which must obtain extra minerals and water from other sources. Sodium, one of the most important elements in animal bodies, is often acquired from salt licks. Thus in some areas the numbers of plant-eating mammals may be limited not by the amount of plant food available but by the number of drinking sites and salt deposits. This applies particularly to elephants, whose requirements for both water and salt are enormous.

Primary consumers which live in a water environment must overcome problems very different from those which face land-living consumers. The phytoplankton on which they feed are dispersed throughout the upper layer of the water, and there is no concentrated food source to compare with leaves or fruit. In the open water of lakes and oceans most primary consumers feed by filtering the plants from the fluid in which they float. Most of the consumers, the zooplankton, are themselves very small and float among the plants. The majority are crustaceans and, to a lesser extent, *tunicates*—that is, animals consisting largely of jelly encasing a gut.

Coastal waters, both fresh and salt, are rich in dissolved minerals and often contain phytoplankton in large amounts. Because of this, many nonfloating animals live on or near the shore, buried in mud or sand or attached to rocks. Examples are shellfish, which suck in water and filter out phytoplankton, and barnacles, which sweep the water with feathery appendages.

But not all consumers of phytoplankton are small. Lesser flamingos feed in warm inland waters, using their tongues as

Below: typical chain of interdependent plants and animals in a hardwood forest environment, using North American species as examples. Predator-prey relationships are shown by the arrows. The relationships shown represent only a fraction of the tens of thousands of similar "chains" that could have been drawn. Yet they demonstrate two basic facts—all food chains begin with plant matter, and the interrelationships grow more complex as the sequence progresses. Such richness forms the foundation of ecological stability.

Right: goose barnacles (*Lepos*), crustaceans which cling tightly to rocks until the tide covers them. Then they open their shells, stick their feet out and sweep them through the water, thereby propelling minute plankton into their mouths.

Below: Caribbean manatee (*Trichechus manatus*), a sea cow with a hairy, creased face. It is a fish-shaped mammal weighing up to 1500 pounds and growing as long as 15 feet. Its skin is hairless, and it has paddlelike forelimbs and a broad, horizontal tail shaped like a shovel. The manatee has no front teeth and its bristly upper lip is divided and mobile so that the two opposing halves can be used as a grasping organ for plucking undersea vegetation, especially eelgrass, or even land plants which might overhang the water.

pumps to suck in water and then pumping it out through the filters that fringe their bills. Some fish also feed mainly on plant plankton—which they trap by driving water over filters called *gill rakers*.

Nor are all water plants planktonic. In shallow water are many larger, nonfloating plants which are eaten by the water equivalent of the grazing animals of the land. Most of these "browsers" are completely aquatic but some are land-based—geese, for instance. Green turtles and Galápagos marine iguanas browse on the submerged fronds of aquatic plants, and mollusks rasp at nonplanktonic vegetation. Large-lipped sea cows, which live in shallow, tropical waters, are unique among mammals, totally adapted to a life of sea grazing.

The next link in the food chain is when primary consumers are themselves eaten by secondary consumers—predatory animals. Until the coming of man some larger herbivores, such as elephants and whales, were rarely killed by predators for food, but most others are hunted by predators of some sort. The owl that eats a shrew that has eaten a predatory insect is a *top carnivore*, obtaining plant-produced food at several removes from photosynthesis. A top carnivore is one that is rarely preyed on by other animals. Fast-swimming fish are typical tertiary consumers of the oceans, joined by such land-based predators as birds and seals. The top carnivores of the sea are the shark and killer whale.

At the upper end of food chains the larger carnivores seldom eat every fragment of the animals they kill. Scavengers—on land hyenas, jackals, and vultures, remoras and pilot fish in the sea—eat what is left over. Size is obviously an important element (for the most part predators are larger than their prey) but there are exceptions. Short-tailed weasels, known as stoats in Britain, kill and eat rabbits several times larger than they are. Generally, however, there is an upper size limit in food chains, and rarely are there more than six links between primary producer and top carnivore.

Food chains may be portrayed diagrammatically in the form of pyramids including several levels, each representing a functional step in the chain. These are called *trophic levels*, and gener-

Links in the Chain

ally the relative numbers of individuals at each level are shown with the producers at the base and the tertiary or higher consumers at the top. Each level of the pyramid must be able to support the level above it on "excess" individuals while maintaining a sufficient breeding stock to insure survival of its species.

An important exception to the basic pyramid model is the *parasitic* element of a food chain. These organisms live in or on the bodies of other living things, obtaining their food in prepared form from the host. They are usually smaller than their victims. Unlike predators, parasites do not eat much of the host's body and rarely kill it. The longer the host survives the more offspring the parasite will be able to produce. Most wild animals and many plants are parasitized. Tapeworms absorb food products made digestible by the enzymes of the host's gut; fleas and leeches suck blood containing most of their necessary nutrient requirements; malaria parasites actually circulate within the bloodstream absorbing food material. Some plants, such as broomrapes (brown, white, or purplish herbs of the genus *Orobanche*, found in the northern hemisphere), suck nutrients from the roots of broom or other host plants. In each case the host expends energy obtaining food both for itself and for the parasite.

Man is involved in many food chains at several trophic levels. He acts as a primary consumer of land plants, a secondary consumer of herbivores such as hoofed mammals, and as a top carnivore of fish. He harbors parasites and sustains scavengers. Many of the other organisms with which he is involved are members of still other food chains. The end result is not a number of separate food chains but a complex food web.

Below: vultures and hyenas feeding from the remains of the carcass of an animal killed by a top carnivore such as a lion. Scavengers are an important link in the food chains of many land and sea environments.

The Role of Decomposition

Right: the common cat flea (*Tenocephalides felis*), shown here on a cat's ear. Fleas are small, wingless insects which live, when adult, as blood-sucking parasites of mammals or birds. Their larvae live on the debris and dirt that accumulate in the lair or nest of the adult's host animal. Cat fleas, finding themselves on a human, will readily bite.

Below: a burying beetle (*Necrophorus vespilloides*), one of various carrion beetles that bury the carcasses of small animals, here a shrew, in which they have deposited their eggs.

But energy cannot accumulate at the top of each food chain—if it did the chains would cease to function for lack of material. It is constantly being disarranged and returned to lower levels through various routes. Bodily wastes and the corpses of dead animals are broken down by decomposition or consumed by other living things, thus releasing their building materials.

Bacteria are the chief agents of decay. They occupy a distinct trophic level, separate from producers and consumers; along with many fungi they are classed as *decomposers*. In feeding they secrete enzymes that break down complex organic matter and release its contents for use at the base of the food chain. They are most abundant wherever wastes and dead matter collect, whether in the soil or the surface water or bed of the sea. Bacteria are so small and so numerous that as many as 455,000 million may inhabit a pound of soil or 570,000 million live in a pint of water. The food of these myriads of decomposers forms the basis for a whole set of further food chains.

Bacteria are themselves consumed, by mites in the soil and many protozoa in both water and earth, while larger animals including man feed on the fruiting bodies of mushrooms and other fungi. But an equally important share of the process of decomposition is performed by *litter animals*—earthworms, millipedes, ants, fly larvae, and termites in the soil, and creatures such as the wormlike pogonophorans in the sea.

After living things die and their bodies are broken down by the action of bacteria and fungi, the basic raw materials of which they are made are released back into their surroundings. Water is the main constituent of animal blood and plant sap, the fluid in which the contents of living cells are dissolved. The human body is about 60 percent water, many plants are 90 percent water, and some water-living animals such as jellyfish contain even more. Water constantly enters and leaves the bodies of living things, lost through evaporation from the respiratory surfaces of land plants (their leaves) and animals (their gills or lungs) and as the medium in which waste products are removed from the body. Water-living organisms usually exchange water continuously with their surroundings, and some freshwater inhabitants even have to pump out some of the excessive amounts of water that flow into their bodies.

Water (H_2O) is made up of hydrogen and oxygen. Together with carbon these are the most common elements in living matter. Along with nitrogen, sulfur, and mineral salts they circulate constantly among plants, animals, the land, sea, and air. To fuel their body processes plants and animals take in oxygen, released by plants from water during the process of photosynthesis, and use it to oxidize carbohydrates, thus releasing energy—a process called *respiration* which is the reverse of photosynthesis.

During respiration every living creature excretes carbon dioxide, usually by the same route through which oxygen enters —the leaves, gills, or lungs. Most of the carbon incorporated into, and retained in, the tissues of living organisms is also eventually returned to circulation when they are broken down, but in the process some carbon becomes trapped and fossilized as peat, coal, and oil.

The fourth most important element in living matter is nitrogen, the most abundant gas in the earth's atmosphere. Though some reaches the soil through lightning, most of it enters food chains by way of nitrogen-fixing bacteria. Nitrates are also broken down through the action of denitrifying bacteria with the result that nitrogen gas is returned to the atmosphere.

Water is constantly carrying matter from the land toward the sea, where there is a steady fall to the bottom. From the accumulation of raw materials on the beds of oceans and lakes some is recirculated back in upwelling currents (caused when cold water sinks and warmer water is pushed up from the lower layers of the

Below: the nitrogen cycle, showing the parts played by various organic and nonorganic factors. About half the dry weight of protoplasm is protein, and about a fifth of this is nitrogen. Nitrogen-fixing bacteria "fix" atmospheric nitrogen by combining it with hydrogen to form ammonia. Later the ammonia is converted to nitrites and nitrates, both of which help legume plants such as clover and alfalfa to synthesize proteins. The main source of available nitrogen in the soil is organic matter itself. When plants and animals die, most of the great store of nitrogen locked up in them returns to the soil. Much is also returned in the form of animal feces and urine. Myriads of microorganisms in the soil then turn the proteins and other complex compounds back into simpler nitrogenous compounds that living plants can absorb.

Denitrification, however, is the reverse of nitrogen fixation. Other bacteria spend their lives breaking down nitrogenous compounds and thereby releasing free nitrogen to the atmosphere. In natural, fertile ecosystems the losses of nitrogen by plant absorption, denitrification, fires, and leaching are more or less balanced by decay and nitrogen fixation. In other ecosystems man removes large quantities of protein by intensive harvesting and returns insufficient nitrogen, in the form of fertilizers or human bodily wastes, to balance the loss.

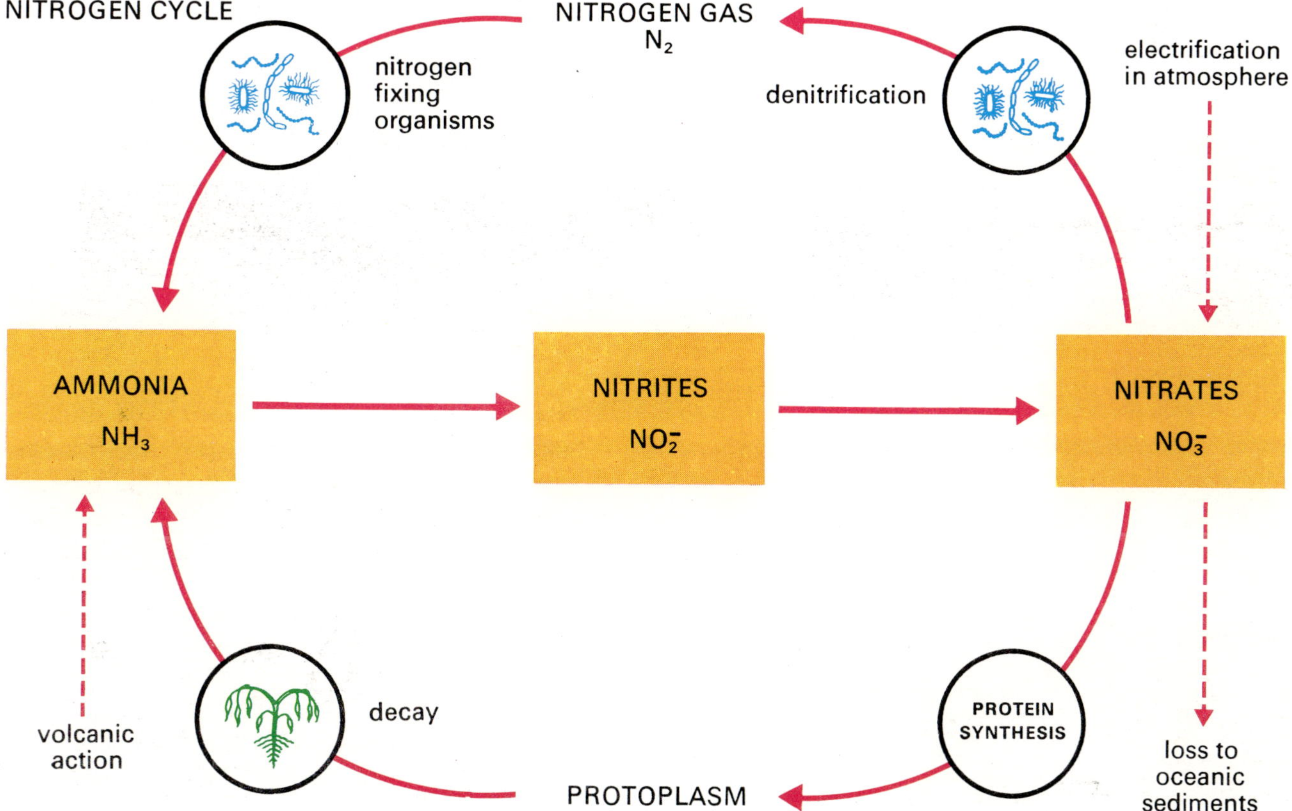

sea). This phenomenon is particularly common in polar waters and warm oceans such as the southeastern Pacific, where the upwelling of mineral-rich currents supports a dense community of plankton and other marine life.

Lack or scarcity of water, that essential component of life, is one of the main characteristics defining one particular large-scale ecosystem—that of warm deserts. A description of this one and of another, completely different ecosystem (a tropical rain forest) will illustrate some of the interlocking factors common to all systems. The effects of the basic physical conditions (*substrate*) and the climate are here simply more apparent.

Warm deserts exist in tropical and subtropical land areas that receive very little rainfall, due to the distribution and movement of atmospheric air masses. The density of plants and animals is much lower than in wetter areas at the same latitude, and plant production is similarly reduced. Deserts are often more barren than polar tundra—sometimes less than one-fiftieth of an ounce of organic material is produced per square yard in a day. Extreme desert conditions can result in less than one-third of an ounce being produced per square yard over an entire year.

Although lack of water was the deciding factor for the formation of most warm deserts, very salty soils combined with only moderately dry conditions can have a similar effect. A rainfall of

Below: tropical rain forest on a hillside near Bukit in Sumatra, an island in Indonesia. In great contrast to the desert's paucity of life is the luxuriance of a tropical rain forest ecosystem. Abundance of water is the key to the difference.

Ecosystem Opposites

less than 6 inches a year in an area of warm climate generally results in desert conditions, while between 6 and 16 inches usually means a semidesert. Heat and lack of clouds cause rapid evaporation of any rain that does fall, which accentuates the arid climate. The rain may also run off so quickly that very little can be used by the plants. The absence of dense vegetation or moist soil allows surface temperatures to rise scorchingly high during the day, and in some places readings of 185°F have been registered. More significantly, the difference between day and night temperatures is usually great—around 35°F—for the desert surface cools rapidly at night due to lack of insulation.

Despite the unfavorable conditions many plants and animals have become adapted to the extreme demands of desert life. The cactus, for instance, a *succulent* (water-storing plant), is a typically well-adapted inhabitant of the American deserts. The root systems of some cacti spread over vast areas near the surface of the soil in order to trap the maximum amount of moisture possible from the infrequent downpours. They store the trapped moisture in their fleshy stems, and evaporation loss is reduced by various characteristic features—the absence of leaves (except the spines which deter browsers); very little branching; a thick waterproof covering layer; and infrequent opening of the *stomata*, apertures in the outer layers, for respiration.

Left: wind-formed sand dunes of the Namib Desert, which fringes the southwestern coast of Africa for a thousand miles. In the north the massive dunes, composed of loose sand, give way to gravel plains occasionally interrupted by flat-topped hills.

Extraordinary Adaptations

Following a rainy spell a huge saguaro cactus, whose root system may have a diameter of 90 feet, may weigh as much as 10 tons, 9 of which are stored water. However, succulents are much less efficient at photosynthesis than are typical broad-leaved trees. Other adaptations for survival include very deep-reaching root systems—for example, in African acacias and the American mesquite the roots may grow as deep as 50 feet below the surface of the soil—and the complete shedding of leaves during the driest seasons so as to lose less water through them, as in some desert shrubs. Indeed, many small desert plants, called *ephemerals*, disappear completely except for their seeds during the driest weather. The seeds survive after the plant itself has died and germinate when more favorable conditions recur. *Boerhavia repens* of the Sahara, related to the South American bougainvillaea, is said to sprout, flower, and seed in only 8–10 days.

Most desert consumers are small and feed only at night, avoiding the heat and dangerously high drying potential of the day air. Small rodents such as jerboas and kangaroo rats are particularly important herbivores. They need not drink but can survive entirely on the water provided in their plant food. This is due partly to a special arrangement of blood vessels in their noses which cuts down water loss during respiration and partly to their characteristic highly concentrated urine, which contains a minimal amount of fluid.

Many herbivorous desert insects are only active, like ephemeral plants, in the wetter times of the year. During the long, hot, dry seasons they survive as eggs or dormant pupae or adults. Others,

Below: giant saguaro cactus (*Carnegiea gigantea*) in the Arizona desert. Most of its bulk is made up of stored water.

Below right: *Onymacris plana*, a beetle of the family Tenebrionidae. There are over 200 species in the Namib Desert, and almost all of them are found nowhere else. Some are active in the heat of the day, when their long legs raise them above the hot sand. Others are nocturnal and burrow during the day. This Tenebrionid beetle drinks condensed fog under a shrub on the sand dunes of the Namib. The condensation also causes the sand to adhere to the insect.

Above: scorpion (*Buthus occitanus*) in the desert. The 650 species of scorpion are found throughout the warm regions of the world but are particularly abundant in deserts, as they are extremely adaptable and can stand fierce heat. They hunt by night and hide by day under rocks or in holes in the sand which they dig with their legs.

like the rodents, are active mainly at night. The next level of the food chain, the secondary consumers, includes carabid beetles, scorpions, and an assortment of lizards and snakes, whose characteristics of waterproof coverings and dry excretion are highly suitable for a desert environment, as is their habit of hiding under rocks, in sand, or in their burrows during the midday hours.

The influence of climate in particular on ecosystems is shown most vividly in the contrast between the desert and the contrasting environment of a tropical rain forest. Annual rainfall often totals more than 120 inches in these equatorial areas of Africa, South America, and the Malaysian-Indonesian archipelago. They enjoy the most stable of climates—constant warm temperatures and some rain in every month.

Vegetation flourishes in this abundant warmth and moisture. Instead of a scattering of small, drought-resistant succulents, giant broad-leaved trees soar to heights of 200 feet. The trees have relatively narrow trunks, often with thin, pale bark, and many are propped up by buttresses developed from their shallow root networks.

Covering the dim forest floor is a sparse growth of herbs. The vegetation forms layers which expose great blocks of photosynthetic tissue to whatever sunlight manages to filter through the uppermost foliage. Above the herb layer, in which arums and gingers are common, is a dense understory of shrubs, and above this rises a layer of small trees. Next, usually between

60 and 100 feet high, is the forest canopy formed from the crowns of large forest trees. From above, these trees give the characteristic visual impression of a rolling green sea. The forest is green year round—some trees produce new leaves for most of the year while others do so during a limited season, but the forest is never bare and there are always some fresh, young leaves.

Compared to the desert vegetation's production of a fraction of an ounce per square yard, the same area in a tropical rain forest has been estimated to produce up to 11 pounds after respiration. Although these forests occupy only about 4 percent of the earth's surface, they account for around 25 percent of its primary production. The ecosystem is rich not only in the quantity of material it maintains but also in the number of plant and animal species it supports, although most species are represented by only a relatively small number of individuals.

It is because of its stable climate that so many species can live side by side in the rain forest. A great number of specialized jobs—ecological niches—exist, and a plant or animal species can become adapted to its particular way of life under specific conditions that remain relatively unchanged over time. Each of a group of closely related species can exist without interruption or undue competition.

In the forest canopy where most of the food supply is produced throughout the year lives a tremendous range of specialized arboreal animals, many of which never descend to the forest floor. For instance, the African forest contains around 30 distinct species of monkey. They act at several levels of the food chain, eating primary producers, primary consumers, and a few secondary consumers, as well as moving at different levels within the forest. They must compete for their food at each level with a huge variety of other species—with birds (hornbills and turacos) for fruit, with caterpillars for leaves, and with insectivorous birds and mammals for insects. Monkeys are themselves preyed on by eagles. The monkeys are just one example of the great variety of life forms that exist in a highly productive ecosystem, flourishing under the most favorable of all land conditions.

Nature is made up of a whole series of intricate biological communities and overlapping ecosystems that interlock to form the earth's *biosphere*. This complex system can be subjected to limited abuse from which it will recover in time, but persistent and large-scale abuse can bring about irreversible changes in the entire network. Man, who is a part of that network, has deceived himself in the past in imagining that he could exploit various ecosystems for his own benefit and suffer no ill effects. We still deceive ourselves in many of our current assumptions about the capacity of an ecosystem to absorb our pollution.

One of the most widespread forms of pollution is the use of pesticides. In the fight against plant parasites and carriers of human disease and in an effort to increase food production a vast armory of chemicals has been developed to kill harmful insects. The use of these chemicals has given rise to a host of unexpected problems. DDT, for example, first used during World War II, is today present in water, soil, and air, stored in human fat, and

The Pollution of Pesticides

drunk in human milk, all thousands of miles away from the nearest spraying activities. Over a billion pounds of DDT have been introduced into the ecosystems of this planet, and in some places up to one-third of the original amount used has been detected in the soil 17 years later.

DDT affects marine life by interfering with the ability of phytoplankton to photosynthesize, thus threatening through the food chain the existence of all the primary and secondary consumers in the sea. In concentrations of as little as 0.01 parts per million (ppm) DDT can inhibit photosynthesis by as much as 80 percent. Due to a process called *biological magnification* the concentration of a nutrient or chemical is increased up the food chain as one organism is consumed by another. After being washed by the rain from fields and forests into rivers and oceans, the practically insoluble material is picked up by the nearest living organisms. As many small creatures are eaten by each larger animal, the concentration builds up to levels where the reproductive cycle of fish-eating birds is disrupted. Species such as the American bald eagle and brown pelican, the Swedish white-tailed eagle, and in Britain the peregrine falcon, sparrowhawk, and golden eagle are all seriously threatened. Even humans are affected—in North America, human milk contains on average over twice the amount of DDT allowed in commercially sold milk (0.05 ppm), and in some states the figure is as high as six times as much. In Australia human milk was found to contain 30 times the maximum permitted dosage. Today there are strict controls on the use of such dangerous persistent insecticides as DDT, but many are still in regular use.

Similarly long-lived are such other commonly used chemical pesticides as the chlorinated hydrocarbons (benzene, hexa-

Below: American C-123 Providers spraying defoliating chemicals over a Vietnamese jungle. By laying bare well over a million acres in the pursuit of narrow short-term political gains, the U.S. government has proclaimed again the dominance of modern man over nature. But what the long-term effects of such actions may be, no one can know as yet.

Above: barn owl (*Tyto alba*) with prey. Top carnivore in a food chain, it is now a threatened species due to susceptibility to insecticide poisoning. Toxic residues persist and are passed along the food chain—first the grain is sprayed with DDT and dieldrin (powerful organic chlorine pesticides); it is then eaten by mice, and they in turn fall victim to owls. Significant quantities of insecticide that thus build up in the owls' livers make them infertile, leading to an inevitable decline in their numbers.

chloride, dieldrin, endrin, and aldrin) and the organophosphates (parathion, Malathion, Azodrin, and phosdrin). They are dangerous additions to the soil, a complex ecosystem in itself which is still comparatively little understood. It supports a numerous and extremely varied organic population—one count, taken in a forest in North Carolina, indicates that around 125 million small invertebrates may inhabit a single acre of soil. And in addition to these mites, worms, and insects are the millions upon millions of bacteria and microscopic plants such as fungi, yeasts, and algae which play a part in keeping the soil fertile and supportive of life. These interconnected systems are all at risk from chemical pesticides, but how and to what extent are still largely unknown.

Unknown Effects

Even less is known about the effects of herbicides on the environment, and we use far more herbicide than we do insecticide. The two basic types include those which cause plants to die by stimulating their internal rhythms, so that they grow themselves to death, and a second group which blocks a crucial step in photosynthesis, causing the affected plant to starve to death. They all act to change the environments—and thus the survival patterns—of the animals which feed on these plants.

Substitutes exist to these dangerous chemical additives. Carbamates and the botanical substances pyrethrum and rotenone (obtained from chrysanthemums and the roots of derris plants respectively) are highly effective yet much less ecologically harmful than the artificial substances. These natural substances, because they are not a source of profit to the powerful agricultural chemicals industry, are not sufficiently promoted, subsidized, or marketed.

Another, more radical, alternative is to shift to a system of agricultural control and organization which would replace previous attempts to exterminate "pests" wholesale. For example, mosquitoes can be reduced in number (something no chemical has succcssfully accomplished) by draining the swamps in which they breed and stocking adjacent waters with appropriate predatory fish. The cotton-eating screwworm of Peru was finally controlled, after all chemical efforts had failed, by neutering males of the species. Although these methods could be used on a wider scale, they are not necessarily cure-alls, and consideration must first be given to the changes in the balance of nature that they themselves undoubtedly cause.

One of the most dramatic examples of modern pollution, showing the devastating effect of industrial processes, was the situation at Minamata Bay, Japan. In the 1950s one of the basic products of a rapidly expanding chemical industry was acetaldehyde, a chemical in great demand for its usefulness in making plastic resins and compounds such as octanol and dioctyl phthalate. The town of Minamata prospered as its thriving chemical plant played its part in trebling the amount of acetaldehyde produced nationally over an eight-year period. The standard production process involved the use of mercuric sulfate as a catalyst and resulted in effluent (outflow) which was allowed to enter the bay.

The first signs of trouble were in 1953 when local cats began to go berserk—some even drowned after throwing themselves into the sea. Soon afterward early symptoms appeared in people—paralyzed hands, dilated pupils, and even mysteriously severe cases of brain damage. This proved to be the beginning of a virtual epidemic in which 45 people eventually died and a further 70, many of them children, were permanently and horribly disabled. At first the chemical plant would admit no liability for what was happening, even when it was established, 10 years later, that the acetaldehyde process resulted in the formation of an extremely toxic compound, methyl mercury, hundreds of tons of which were accumulating in the waters of Minamata Bay and in the bodies of its fish. The company continued to operate in the same way, and court actions brought against it were bitterly resisted. Ultimately the company was forced to pay compensation to the families of victims of what was established to be mercury

Is There a Safe Exposure Level?

Right: Chisso plastics plant in Minamata, the Japanese seaport which has become a byword of environmental tragedy. Thousands were poisoned and dozens killed through the release of mercury from this factory in a deathly process that only came to light in 1953.

poisoning, and to acknowledge its responsibility, but the effects of the pollution persisted. By the mid-1970s some 500 cases had been officially verified in the bay area, and unofficial estimates ranged from 1000 to 10,000 people still at risk, since a population of 200,000 people had lived on the fish caught in the contaminated waters of Minamata Bay.

Another material that threatens the health of the public, as well as of those who work in direct contact with it, is asbestos. Processed on a commercial scale for nearly a century, it is used for jobs as varied as insulating pipes and lining ironing boards. Its most common use is in vehicle braking systems, so that accompanying the meteoric rise in production of trains, cars, trucks, and planes the consumption of asbestos has grown from a total worldwide figure of 30,000 tons in 1910 to well over 4 million tons a year in the 1970s.

But even as long ago as the 1890s, during the first five years after the opening of an asbestos mill in France, 50 workers there died of disease that could be connected with the fibrous materials with which they worked. In the early years of this century asbestos dust was clearly identified as the toxic agent of asbestosis, an incurable cancerous condition of the membrane surrounding the lungs, which has proved increasingly prevalent among workers in the industry as well as among people who are in contact with the manufactured material: builders, for instance, use asbestos cement or asbestos wallboard, and mechanics operate brake-testing equipment.

But asbestosis is a health hazard not limited to occupational risk. The average car goes through four sets of brake linings and two clutch linings during its serviceable life, and while these are wearing out they give off asbestos particles into the environment which are then inhaled by everyone. An eminent pathologist has claimed that asbestos dust will soon rival cigarettes as a cause of lung cancer. Yet it is still used: soft drinks are filtered through it, surgeons use the powder to speed up healing of wounds, and children are sometimes given it mixed with water to play with—and a mere teaspoonful can be fatal. The victim may not die for as long as 20 years after exposure to the fibers, but the eventual result is a certainty.

Potentially the most hazardous of all substances, however, is plutonium, the artificial radioactive element produced when uranium has been processed in a nuclear reactor. Itself a fuel for more advanced reactors, it is possibly the most poisonous substance known to man. A piece the size of an orange contains enough cancer-causing material to afflict every person on earth with leukemia, and a thousandth of a gram, smaller than the point of a pin, is enough to kill a man. Yet nuclear plants have been dealing with it in 5- to 15-ton quantities for years. In both the United States and Britain, both of which operate advanced nuclear-power programs, anxiety is increasing regarding both the safety of workers in nuclear plants and the exposure of the general public to radiation emitted by plutonium waste by-products.

One effect of handling plutonium is now thought to be that the chances of contracting cancer of the bone marrow are at least 20 times higher than usual. Many scientists believe that there is not even a safe maximum level of exposure to radiation, as is currently set by government agencies, and that any exposure at all should be considered unacceptable. There is also increasing concern in several countries about the disposal of radioactive wastes, which must be kept from entering the living environment for hundreds or even thousands of years. The problem of securing the wastes so that there is no danger of leakage is far from solved. In the short term they can be kept in relative safety (barring deliberate sabotage or theft), but in the longer term we may be putting generations and cultures yet unborn at great risk.

Below: the Windscale nuclear power station in Windscale, Cumberland, in northwestern England. It is a fast-breeder nuclear reactor center producing plutonium from uranium. Public inquiries into its future sponsored by the British government have led to widespread publicity over questions of safety for both humans and the environment.

Inadvertent Destruction?

Radioactive material passes easily through the human body and affects the nuclei vital for cell multiplication. A substance such as strontium-90, which is stored in bone marrow, can kill cells outright and has been linked to diseases such as leukemia. It is an artificial isotope formed by nuclear explosions and is considered the most dangerous constituent of fallout. Strontium-90 is absorbed by grass and when eaten by cows turns up in meat and milk, replacing some of their calcium content. Absorbed by humans (especially children) it ultimately becomes concentrated in their bones and teeth. It is radioactive for an extremely long time—28 years must pass before it has decayed to even half its original strength.

Even before the advent of modern technology man was responsible for creating deserts and barren wastes in some parts of the world. The period from about 5000 B.C. to 200 A.D., which produced some of the great civilizations—Sumeria, Babylonia, Assyria, Phoenicia, Egypt, Greece, and Rome—was also a time when agriculture was establishing itself. For instance, the area that is now Iran and Iraq was once covered with forests and grasslands, watered by the Tigris and Euphrates rivers, and herds of domesticated cattle, sheep, and goats could be supported in the area. From around 6000 B.C. the forests were cut down to provide more land for the increasing human and livestock populations, and by 4000 B.C. timber was being used for building cities. With the passage of time irrigation schemes, at first both useful and efficient, resulted in the accumulation of silt so that after 3000 B.C. the Persian Gulf had become filled in for a total of 180 miles from its original position. Wars contributed to the devastation, and today most of the land is desert.

Below: the American gray squirrel (*Sciurus carolinensis*), a species of rodent sometimes called a tree rat. Specimens that escaped from two British zoos later multiplied and killed off almost all the native red squirrels (*S. vulgaris*).

Right: man's role in animal dispersal, which includes regulation and control in addition to the introduction of new species. The denuded land on the left shows the result of the increase in the rabbit population of Australia, which was up to over 100 million by 1950. The same land is shown on the right after myxomatosis had been artificially introduced as a means of control, causing a population reduction of 75 percent.

The Sahara desert, mainly a result of major climatic shifts over long periods of geological time, has been extended by man in recent centuries. The cedar forests that once covered North Africa have disappeared, primarily due to Roman exploitation, and now the barren hillsides are almost denuded of vegetation and most of the once-fertile soil has been lost.

The fragile workings of an ecosystem have also been upset by the introduction of non-native animals. Australia, for example, has received both invited and stowaway animal and plant guests, one of the earliest being the introduction of the European rabbit in 1859. From the 12 pairs originally released in Victoria the population exploded, and six years later during one extermination campaign 20,000 were shot. By the turn of the century several hundred million were found covering over two-thirds of the continent. The reasons were an absence of natural predators (apart from the dingo dog) combined with the rabbit's well-known reproduction rate. The introduction in 1868 of the European fox had disastrous effects on the indigenous bird and marsupial populations but did little to halt the spread of the rabbits. They destroyed by overgrazing the range and pastureland farmers wanted for their sheep, and neither poison, fencing, nor shooting reduced their numbers sufficiently. In 1950 a myxoma virus was introduced into local rabbit populations. The highly infectious, fatal disease produced fever, swelling, and localized mucous tumors on the skin. Spread by mosquitoes, myxomatosis rapidly became an epidemic, but although at first between 80 and 90 percent of the rabbits in a local population would be killed, their numbers later increased again due to the breeding of the survivors. With succeeding epidemics the numbers of victims decreased due to the build-up of immunity, and today the numbers are again very high, causing a nuisance and economic problem. Much the same sequence of events took place in New Zealand, where the first rabbits were welcomed as pets, for sport, and for the value of their skins.

One of the most famous examples of accidental importation of pests into an area to which they are not native was discovered in 1930 by an entomologist in Natal, on the northeastern coast of Brazil. He found a specimen of the malarial mosquito *Anopheles gambiae*, a native of Africa and hitherto unknown in Brazil. Within a year *A. gambiae* had spread 115 miles along the coast, and by 1938 its population had become so large that it set off a malaria epidemic that killed 14,000 Brazilians in only six months. The Brazilian government managed to eradicate the mosquito successfully in 1940, but, although *A. gambiae* was undoubtedly an importation, exactly how it was introduced has never been known for certain.

In the last decade or so a whole series of related reports, statements, and new attitudes has emerged that together may be the first deep rumble of an approaching revolution in man's relationship with this planet. In industrialized societies there are growing pressures for less noise, cleaner rivers and air, safer food, and less wasteful industries. However small and limited these decisions may seem in relation to world ecological problems as a whole, they have a significance larger than can be measured by their purely local impact. The limited store of fossil fuel and

Below: poisoned fish, indirect casualties of Britain's well-intentioned Clean Air Act passed in 1956. The law led to the establishment of many large factories run on smokeless fuel, and their highly toxic wastes then began to seep or be dumped into nearby rivers. One polluting process is often superseded by another in this way, as humans tend to think in linear fashion (action A solves problem B) instead of cyclically, which is the way nature works.

Man's Role in a Global Scheme

difficulties in producing energy are now being clearly realized for the first time, but it would be an overstatement to suggest that such pressures have changed the course of technological development. Mankind has not yet reached the stage of thinking in terms of one earth, vulnerable and tightly stretched life support systems, or of eventual equilibrium. But some of the worst ecological wounds are being recognized and patched up at last, and this is a hopeful sign. The future of mankind depends absolutely on the acceptance of a new scale of values to which existing (especially Western) institutions, economies, and cultures are inherently opposed. Predatory and short-sighted, man is locked in self-destructive systems of his own making. The situation is complicated by the obvious difference in wealth (and ability to pollute) between industrialized and developing nations. The Third World, paradoxically, cannot afford the "luxury" of non-pollutants in its monumental task of feeding and employing its people.

The attitude of industrial societies to the dangers of lead, a highly potent nerve poison, is a good illustration of the present state of affairs. Wherever lead is smelted or in some other way involved in industrial processes, such as the casting of pipes or of lead alloy components, considerable attention has been given to the control of fumes. The workers are kept under continual medical observation to make sure their lead intake remains below the level of obvious clinical poisoning. Using this specialized group of people as a control, guidelines are laid down arbitrarily for what qualifies as lead poisoning. These criteria are not necessarily applicable to whole populations, in which individuals vary enormously in their sensitivity to particular poisons. They also for the most part take no account of the behavioral effects of lead or of its ability to cross the placental barrier between mother and unborn child and thus to affect the developing fetus in its most vulnerable stages. Nor do they take note of the general and gradual build-up of lead in the overall environment, the biosphere. The guidelines also exclude mention of the lead-tolerance of children and do not apply to the organic compounds of lead that now make up a large part of lead contamination in urban societies.

Lead exists in low concentrations almost everywhere in the earth's crust, and through natural processes about 150,000 tons a year return eventually to sediments. Presently man is mobilizing about 5 million tons (and the rate is growing) each year, so that the burden on various life cycles is 30 times greater than normal. Car exhaust fumes account for 400,000 tons of lead a year in Europe and the United States alone, and in this form they could hardly be better designed for rapid absorption into living mammals' systems through their lungs. Lead has been found to be damaging to biological processes wherever it has been studied, and so, on the one hand, society treats industrial use of lead with the most stringent controls (including space suits). But, on the other hand, a wholly unprotected public is sprayed with these same compounds in vast quantities. Only about 5 percent of the lead mined, refined, and processed is industrially recycled, so that the rest, by one route or another, eventually enters natural ecosystems and builds up in plants and animals—including man.

Lead is particularly poisonous to animals at the top of food chains, which have the most highly developed nervous systems, and the effects—ranging from acute poisoning when lead stored in the tissues is released, to serious and irreversible brain damage and behavioral changes—are already becoming evident in a high proportion of the urban child population.

So it could be said that the major sources of modern pollution are not the effluent pipes and smokestacks of a callous industry, controllable by the simple addition of emission control devices. They are the very technologies themselves that produce both the smoke and the devices, part of the whole structure of modern industrial society. The damage that has been done accumulates inexorably and cannot be cured by slight measures—if we ceased using DDT tomorrow, for example, it would still be carried in rain falling worldwide a century from now. Any activity alien to the cyclical biological systems that have evolved on earth is likely to be damaging and is certainly disruptive, and therefore almost every aspect of industrial technology is bound to provoke an ecological backlash. The processes of modern man are straight and open-ended, whereas nature's are almost entirely cyclical.

Man has not yet become fully aware of his role as a participant in a global scheme. But perhaps that awareness will arrive in time for us to make the drastic alterations necessary in order to save our planet and its inhabitants, including ourselves. But time is unarguably running out, and the strain placed on the earth's ecological cycles may soon be irreversible. For there is one thing that man cannot do. In spite of all his accomplishments in terms of "tinkering" with nature, he is unable to recreate those most intricate of its marvels—the fragile networks of life.

Left: reflectors of the solar furnace at Odeillo in the French Pyrenees mountains. The central furnace can reach temperatures of 6000°F. This modern technological answer to the so-called energy crisis underlines the fact that the sun, source of all the earth's energy in one form or another, actually radiates sufficient power to replace all man's current energy sources with power to spare. Even in the cloudy latitudes of Great Britain the average amount of solar energy penetrating the atmosphere and arriving at ground level is over 100 watts per square yard, or 400 times the present output of all its existing power stations combined. If man could replace his very restricted set of energy sources—fossil fuels and nuclear materials, which are fast running out—with collecting the sun's energy (even with only partial efficiency) on a large scale, the resulting power would exceed the most extravagant current usage without pollution, dangerous radioactivity, or imminent risk of exhausting the supply.

Index

Figures in italic refer to illustrations.

Picture Credits

Key to picture positions: (T) top; (C) center; (B) bottom; and in combinations, e.g. (TR) top right; (BL) bottom left.

Aldus Archives 33(T), 46, 48, 63, 104, 164(R); Cathy Jarman, *Atlas of Animal Migration*, Aldus Books Limited, London, 1972 139(B), 140, 142, 144, 150(T), 151, 153; © Aldus Books 36, 37(B), 38(T), 45, 57(T), 69, (Gordon Cramp) 18–19(B), 20, (Design Practitioners) 56, (Geoffrey Drury) 209(B), (Edward Poulton) 237, (Edward Poulton, after G. A. Eiby) 19(T), (Maurice Sweeney) 209(T), (Peter Warner) 233, (Maurice Wilson) 72, 81, (Sidney W. Woods) 21(B); American Museum of Natural History 77, 90; Dr. Peter Andrews 112(B); Heather Angel/Biofotos 244; Ardea (P. J. Green) 78, (Peter Green) 79, 86, (Arthur Hayward) 71, (Richard Waller) 41, (Tom Willock) 11(B); Bob Davis/Aspect 246; Associated Press 106(B); Australian News and Information Bureau, Sydney 248(R); Donald Baird, Princeton University Museum 88(T), 112(T); Bavaria-Verlag 24(B); The Bettmann Archive Inc. 146; British Museum/Photo John Freeman © Aldus Books 29; British Museum (Natural History) 28(T), 60, 62, 66(B), 73(R), 83(T), 106(T), (Dr. Allan Charig) 97, (Photo Mike Busselle © Aldus Books) 165, (Photos John Webb © Aldus Books) 35, 82–83(B); Photos by Ralph Buchsbaum 87(L), 88(B); Zdenek Burian 67, 70; Zdenek Burian, *The Dawn of Man*, by courtesy of his publishers, Artia, Praha 194; Photo Colin G. Butler, F.R.P.S. 215(T); Camera Press 247; Restoration by Robert T. Barker, in D. A. Russell, *Canadian Geographical Journal* 87, pp. 4–11 74(L); Canadian Press Picture Service 38(B); Photos J.-L. Charmet 66(T), 131, 143; Cleveland Museum of Natural History 110; Bruce Coleman Ltd. 101, 121, 177(R), 191, 195(B), 198, 200, 201(T), 202–204(T), 206, 208, 232, 234(B), (Des Bartlett) 100, (Jen and Des Bartlett) 108(T), (Mark Boulton) 113(B), (John and Sue Brownlie) 30, (Jane Burton) 54(R), 199, (R. I. M. Campbell) 113(T), (Francisco Erize) 52(T), 54(L), (M. P. Harris) 44(T), (David Hughes) 84, (John Markham) 99, 175(B), 230, 248(L), (G. D. Plage) 11(T), (Alan Root) 44(B), 96, (Werner Stoy) 8; Arthur Lakes Library, Colorado School of Mines 65(B); Mary Fisher/Colorific! 249; Matthieu Ricard, *The Mystery of Animal Migration*, Constable & Co. Ltd., London, 1969 130, 150(B); Cooper-Bridgeman Library 123, 205; Culver Pictures 64(T); Robert J. Ellison/*Daily Telegraph* Colour Library 243; Robert Descharnes, Paris 26; Dr. Horace E. Dobbs 127; Down House/Photo Eileen Tweedy © Aldus Books 50; Edward Poulton © Aldus Books, after *Knaurs Jugendlexikon*, Droemersche Verlagsanstalt, München/Zürich 31(B); Edimages 10; Marcel Brion, *The Glory and the Grief*, Elek Books Ltd., London 15(B); Mary Evans Picture Library 17(T), 52(B); FPG 73(L), 105(B); Field Museum of Natural History/Paintings by Charles R. Knight 75–76; Giraudon 14; Professor M. F. Glaessner, University of Adelaide 32(T); High Commissioner for New Zealand 125; Peter Hill 167; Historical Pictures Service 28(B); Michael Holford Library photos 195(T), 196(L), 197, 204(B); I.B.A. 31(T), 57(B); Institut Royal des Sciences Naturelles de Belgique, Départment Paléontologie 74(R); after Gustav Kramer, *Eine neue Methode zur Erforschung der Zugorientierung und die bischer damit erzielten Ergebnisse*, Proc. 10th International Ornithological Congress, 1950, pp. 269–80, Uppsala 134; Jacana 147, 154, 157, 158, 166, 181(B), 210, 222, 224, 235, 241, (Moiton) 176, (Philippe Varin) 98, (Ziesler) 178; Keystone 251; Laboratoire d'Anatomie Comparée du Muséum, Muséum National d'Histoire Naturelle 92; Photo Lilly and Miller 119(T); The Mansell Collection, London 16, 49(B), 53(T), 55, 95, 107(B); Marine Studios, Marineland, Florida 116, 118, 119(B), 124; Collection Musée de l'Homme 108(B), (Cliché José Oster) 102; National Maritime Museum, Greenwich/Photo © George Rainbird Ltd. 1969 51; National Museums of Kenya 109, 111; National Portrait Gallery, London 49(T); Natural History Photographic Agency 2–3, 135–136, 138(B), 145, 156, 159, 161, 177(L), 179, 184–185, 201(B), 207(B), 212, 214, 216–217, 218(T), 220–221, 226–227, 231, 236, 238, 240, (Stephen Dalton) 180, 183, (Peter Johnson) 181(T); *Nature* 39; Naturfoto 225, (Elvig Hansen) 213, (Gerth Hansen) 187(BL), (Karl Holgard) 187(BR), (I. Mohl Madsen) 187(T), (Ulla Schmidt) 168; Oxford Scientific Films 141, 162, 169–170, 175(T), 182, 188, 190, 215(B), 218–219(B), 234(T), (Animals Animals) 163; Property of Famille Lhoeste, © Patrimoine de l'Institut Royal des Sciences Naturelles de Belgique 105(T); Courtesy The Peabody Museum of Natural History, Yale University 64(B), 65(T); Photoresources 114; Hervé Chaumeton/Photos Scientifiques 173–174, 186, 192; Photri 22(L); Picturepoint, London 15(T), 23, 58; Pitch (M. Sester) 172(R), (G. Vienne) 59; Popperfoto 107(T); *Radio Times* Hulton Picture Library 47, 53(B); Phil Roach/Rex Features 18(T); G. R. Roberts, Nelson, New Zealand 94; Dr. Edward S. Ross 239; *Science Journal*, London 34; Scientific American Inc. All rights reserved (after E. G. F. Sauer, *Celestial Navigation by Birds*, August, 1958) 135(B), (Shirley Parfitt © Aldus Books, after *Scientific American*, 1962) 137, (S. T. Emlen, August, 1975) 134, 135(T), (June, 1975. Courtesy Sarah Landry and E. O. Wilson) 223; Margaret M. Smith, J. L. B. Smith, Institute of Ichthyology, Rhodes University, Grahamstown, South Africa 91(R); Smithsonian Institution. U.S. National Museum of Natural History 87(R); Solarfilma 37; Space Air Photos 25; Tom Stack & Associates 120, 128, 138–139(T), 148–149, 152(L), 160, 172(T)(BL), 189, 207(T), (Keith Gillett) 42, (Tom Stack) 89, 93, (Donald E. Trinko) 117; John Topham Picture Library 22(R); UPI (Compix) 24(T); United States Information Service 21(T); Universität Tubingen, Institut und Museum Für Geologie und Paläontologie 68; Tabley Collection, University of Manchester 12–13; after p. 29 of Wolfgang Wickler, *Mimicry in Plants and Animals*, Weidenfeld and Nicolson Ltd., London, 1968 164(L); Woodmansterne Ltd./John E. Guest 17(B); Rainer Zangerl, *Schweizerische Palaeontologische Abhandlungen*, Vol. 73, 1958 32(B); Zefa 228.